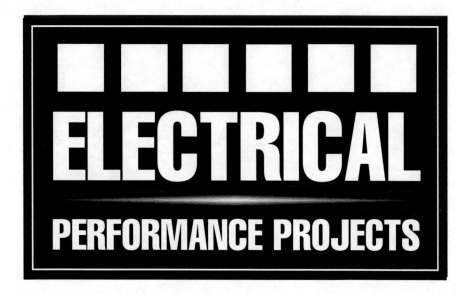

ELECTRICAL
PERFORMANCE PROJECTS

Upper Saddle River, New Jersey
Columbus, Ohio

contren®
Learning Series

10 9 8 7 6 5 4 3 2 1
ISBN 0-13-228216-X

INTRODUCTION

Each module in the Electrical Curricula is designed to help students learn a specific skill. The Performance Projects provide the students with the opportunity to practice these skills as they study the modules. Some of the projects are pencil and paper exercises, such as load calculations, but most are hands-on activities that reflect situations the electrician will face on the job. Every effort has been made in the development of Performance Projects to have them mirror the learning objectives and performance tasks for each module. Most of the projects can be used to fully or partially satisfy the performance objectives of the modules they support.

It is our hope that these Performance Projects will not only provide valuable learning experiences that reinforce the four levels of electrical training, but will also inspire the student to expand their level of skills and knowledge through continual practice and application of these skills.

TABLE OF CONTENTS

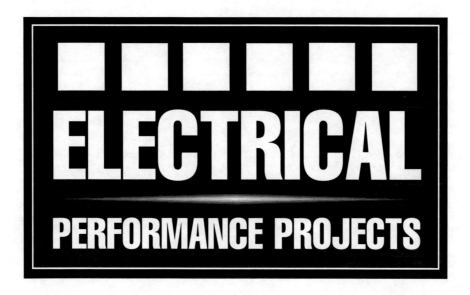

MODULE 26301-05

Projects:

- ■ 1-1 Calculate Load and Determine Feeder Size for a Three-Phase Commercial Kitchen Containing Multiple Pieces of Equipment

- ■ 1-2 Calculate Load and Determine Feeder Size for a Three-Phase Feeder Circuit Supplying Multiple Motors

ELECTRICAL

PERFORMANCE PROJECT

Name_____ Date _____

PROJECT OVERVIEW

Before any electrical service can be installed, a load calculation must be performed to determine the connected load and the demand factor on the electrical service, as well as the feeder sizes. The nature of the load is a major factor in calculating the demand that various loads place on the electrical service. Loads that are considered continuous-use, meaning that they are expected to operate continuously for three hours or more, typically impose a greater demand than intermittent loads.

If a commercial kitchen has no more than two pieces of commercial kitchen equipment, including any electric cooking equipment, dishwasher booster heaters, water heaters, or other kitchen equipment, the nameplate load ratings are added together and the demand factor on the service is 100 percent. However, the total demand factor for three or more pieces of commercial kitchen equipment connected to the same service can be calculated at less than 100 percent, according to the *National Electrical Code®*, *Section 220.56* and *Table 220.56*.

This project asks the trainee to calculate the overall load of multiple pieces of three-phase commercial equipment in volt-ampere units and in ampere units based on a three-phase service of 208 volts.

OBJECTIVES

This performance project supports the following learning objectives listed in Module 26301-05:

- Calculate loads for single-phase and three-phase branch circuits (*Objective 1*).
- Calculate ampacity for single-phase and three-phase loads (*Objective 4*).
- Use load calculations to determine branch circuit conductor sizes (*Objective 5*).

PERFORMANCE TASKS

There are no performance tasks associated with this module.

MATERIALS REQUIRED

- None

TOOLS AND EQUIPMENT REQUIRED

- Calculator
- Pencils and paper

REFERENCE MATERIALS AND LEARNING RESOURCES

- Module 26301-05, Sections 1.0.0, 1.1.0, 1.3.0, and 8.0.0
- *NEC Section 220.56* and *Table 220.56*

NOTES TO TRAINEE

- Round off all answers to two decimal places.
- Remember that you are calculating a three-phase service.
- Consider any load listed in watts as a volt-ampere load.

NOTES TO INSTRUCTOR

- The solution for this project can be found at the end of this project.

PROCEDURE

This performance project requires you to calculate a commercial kitchen load, determining the total nameplate load in volt-amperes, demand factor load, and feeder current rating.

1. Refer to Figure 1. Determine the total nameplate load in volt-amperes.
2. Apply the demand factor from *NEC Table 220.56* to the total VA load and write your answer directly on Figure 1 in the space provided.
3. Calculate the feeder current rating and write your answer directly on Figure 1 in the space provided.

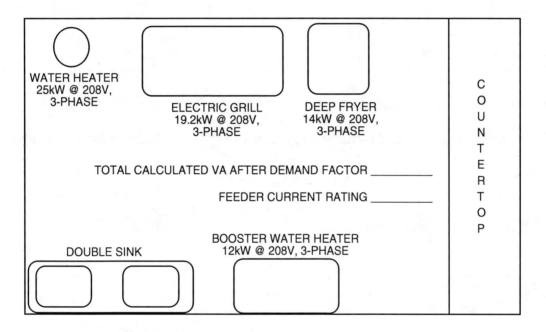

Figure 1 ■ Commercial Kitchen Electrical Loads

Module 26301-05
Project 1-1: Calculate Load and Determine
Feeder Size for a Three-Phase Commercial Kitchen
Containing Multiple Pieces of Equipment

SOLUTION

Total loads:

Water heater	25,000 VA
Grill	19,200 VA
Deep fryer	14,000 VA
Booster water heater	12,000 VA
Total VA	70,200 VA

Demand factor from *NEC Table 220.56*:   80 percent

Calculated load after demand factor of 80 percent: $70,200 \times .80 = 56,160$ VA

Amperage: $56,160 \text{VA} \div (208 \times 1.732)$
$56,160 \text{VA} \div 360.26$

Feeder minimum amperage rating: 155.89 amperes

Name_____ Date _____

PROJECT OVERVIEW

The *National Electrical Code®* states that you must not use the actual current rating marked on the motor nameplates when calculating the ampacity of conductors, but must use the full-load current tables located in *NEC Article 430*.

The *NEC®* also states that when conductors are supplying several motors, or a motor along with other loads, the conductor must meet the following conditions: an ampacity of not less than 125 percent of the full-load current rating of the highest rated motor plus the sum of the full-load current ratings of all the other motors being supplied by the same conductors, plus any other ampacity ratings required for other non-motor loads (if any).

In order to complete this project successfully, the trainee must determine the individual full-load current ratings of each motor by referencing the correct table in the *NEC®*, and must then calculate the minimum feeder conductor rating in amperage according to the requirements of the *NEC®*.

OBJECTIVES

This performance project supports the following learning objectives listed in Module 26301-05:

- Calculate loads for single-phase and three-phase branch circuits (*Objective 1*).
- Calculate ampacity for single-phase and three-phase loads (*Objective 4*).
- Use load calculations to determine branch circuit conductor sizes (*Objective 5*).
- Select branch circuit conductors and overcurrent protection devices for electric heat, air conditioning equipment, motors, and welders (*Objective 7*).

PERFORMANCE TASKS

- There are no performance tasks associated with this module.

MATERIALS REQUIRED

- None

TOOLS AND EQUIPMENT REQUIRED

- Calculator
- Pencils and paper

REFERENCE MATERIALS AND LEARNING RESOURCES

■ Module 26301-05, Sections 1.0.0, 1.1.0, 1.3.0, and 12.0.0
■ *NEC Sections 430.5(1), 430.24, and Table 430.250*

NOTES TO TRAINEE

■ Round off all answers to two decimal places.
■ Remember that you are calculating a three-phase service.
■ Use the full-load current ratings located in *NEC Table 430.250*.
■ All motors are assumed to be three-phase, 460 VAC, induction-type squirrel cage motors.

NOTES TO INSTRUCTOR

■ The solution for this project can be found at the end of this project.

PROCEDURE

This performance project requires you to calculate the total load for multiple motors supplied by a single feeder circuit, and determine the minimum feeder current rating.

1. Refer to Figure 1.
2. Look up and write down the full-load amperage for each motor shown directly on Figure 1 in the spaces provided.
3. Determine the minimum feeder conductor size in amperes of a feeder circuit supplying all of these motors, according to the requirements located in *NEC Section 430.24*. Enter your answer directly on Figure 1 in the space provided.
4. Have your instructor check your work.

SUPPLEMENT

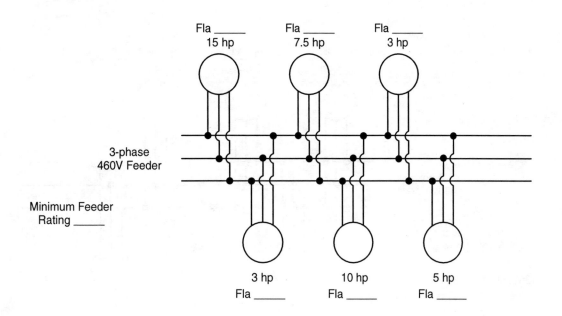

Figure 1 ■ Three-Phase Feeder Circuit

Module 26301-05
Project 1-2: Calculate Load and Determine
Feeder Size for a Three-Phase Feeder Circuit
Supplying Multiple Motors

SOLUTION

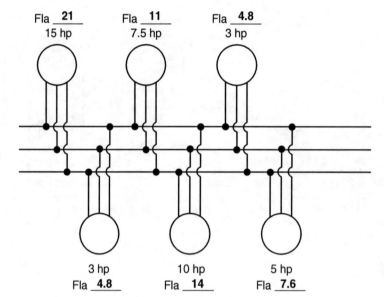

3-phase
460V Feeder

Minimum Feeder
Rating **68.45 Amperes**

125% of 21 = 26.25
100% of 11 = 11.0
100% of 4.8 = 4.8
100% of 4.8 = 4.8
100% of 14 = 14.0
100% of 7.6 = 7.6
_____
Total Amperage = 68.45

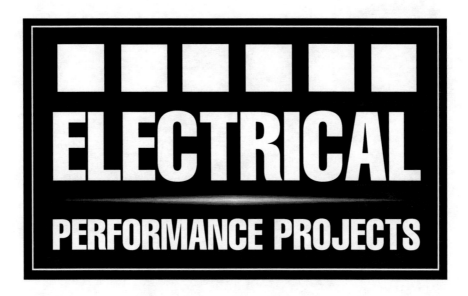

MODULE 26302-05

Projects:

- ■ **2-1** Select the Minimum Size THHN Copper Conductor Size for a Given Load

- ■ **2-2** Measure Actual Resistances of Lengths of Wire and Compare Measurements to Calculated Values Based on Resistance Values in *NEC Table 8*

ELECTRICAL

PERFORMANCE PROJECT

Name_____ Date _____

PROJECT OVERVIEW

When calculating wire size, electrical designers and electricians must consider other factors in addition to the amount of current that will flow through conductors. Examples of these factors include conductor length, ambient temperature, and the number of conductors that will be installed in the same conduit.

Two important conductor relationships must be considered: as conductors increase in length, resistance increases; and as conductors increase in size, resistance decreases. Earlier training modules taught us that voltage is dropped across a resistor and that the amount of voltage dropped depends on the amount of resistance present. Any conductor, regardless of its length, contains some resistance. Tables in the annex section of the *National Electrical Code®* list resistance per 1,000 feet of conductor based on the AWG size of the conductor. If the conductor is extremely long, the voltage drop from one end of the conductor to the other can affect the operation of the load connected to the circuit. Minimizing voltage drop in a lengthy circuit can be accomplished by increasing the size of the conductor. The *NEC®* does not regulate voltage drop, but does recommend a maximum feeder circuit voltage drop of 5 percent.

Conductor amperage ratings for a maximum of three current-carrying conductors installed in conduit in an ambient temperature of 86°F are listed in *NEC Table 310.16*. If the ambient temperature exceeds 86°F, or the number of current-carrying conductors exceeds three in a conduit, the ampacity of the conductor must be derated (reduced).

In this project, the trainee will select the minimum THHN copper conductor size for a given load based on derating factors and voltage drop associated with a specified feeder conductor installation.

OBJECTIVES

This performance project supports the following learning objectives listed in Module 26302-05:

- Select electrical conductors for specific applications (*Objective 1*).
- Calculate voltage drop in both single-phase and three-phase applications (*Objective 2*).
- Size conductors for the load (*Objective 5*).
- Derate conductors for fill, temperature, and voltage drop (*Objective 6*).
- Select conductors for various temperature ranges and atmospheres (*Objective 7*).

PERFORMANCE TASKS

- There are no performance tasks associated with this module.

MATERIALS REQUIRED

■ None

TOOLS AND EQUIPMENT REQUIRED

■ Pencils and paper
■ Calculator
■ Latest edition of the *National Electrical Code®*

REFERENCE MATERIALS AND LEARNING RESOURCES

■ Module 26302-05, Sections 4.0.0, 4.2.0, 4.3.0, 4.4.0, and 4.5.0
■ *NEC Sections 210.19(A) FPN No. 4 and 215.2(A)(3) FPN No. 2*
■ *NEC Tables 8, 310.15(B)(2)(a), and 310.16*

NOTES TO TRAINEE

■ No safety equipment is required for this project unless the environment in which the project is completed requires safety equipment.
■ Conductor type is THHN stranded copper wire.
■ This project assumes that voltage drop is being calculated for only two of the six current-carrying conductors in the conduit, but you must still derate for the total number of current-carrying conductors and the ambient temperature in the equipment room.
■ Voltage drop percentage may not exceed 5 percent of supply voltage.
■ If voltage drop percentage exceeds 5 percent of supply voltage, you must increase the size of the conductors until the voltage drop percentage is less than 5 percent of the supply voltage.
■ All voltage drop calculations assume the use of copper conductors with a K value (constant) of 12.9, which represents the resistance in ohms per mil foot.
■ The circuit is single-phase, 240-volt feeder circuit.
■ Use the voltage drop formula:

$$VD = \frac{2 \times L \times K \times I}{CM}$$

Where:
VD = voltage drop
L = length of run
K = constant of 12.9
I = current
CM = circular mil dimension of the conductor
2 = the length times two

■ Remember that VD represents the actual voltage that is dropped and not the percentage. In order to calculate the percentage of VD, use the following formula:

%VD = (VD ÷ supply voltage) × 100

NOTES TO INSTRUCTOR

■ This is a feeder circuit installation and the voltage drop may not exceed 5 percent.

■ This project assumes that we are calculating voltage drop for only two of the six current-carrying conductors in the conduit, but we must still derate for the total number of current-carrying conductors in the conduit and the ambient temperature in the equipment room.

■ Remember that VD represents the actual voltage dropped and not the percentage. In order to calculate the percentage, use the following formula:

%VD = (VD ÷ supply voltage) × 100.

■ The solution for this project can be found at the end of this project

PROCEDURE

This performance project requires you to select the minimum size THHN copper conductor that can be installed based on given criteria.

1. Refer to Figure 1. Determine the minimum size THHN copper conductor that can be used in a 40-ampere feeder circuit.
2. Derate the ampacity of the conductor size found in Step 1 for the ambient temperature and the number of current-carrying conductors in the raceway, as shown in Figure 1.
3. If the derated amperage is less than 40 amperes, select the next larger size conductor and derate its ampacity.
4. Continue this process until the derated ampacity is 40 amperes or greater.
5. Locate the circular mil dimension for the AWG size conductor determined in Step 4.
6. Calculate the voltage drop using the following formula:

$$VD = \frac{2 \times L \times K \times I}{CM}$$

7. If the voltage drop calculated exceeds 5 percent of the supply voltage, increase the size of the conductor to the next higher AWG number and recalculate the voltage drop using the larger conductor's circular mil dimension.

Remember that VD represents the actual voltage that is dropped and not the percentage. In order to calculate the percentage, use the following formula:

%VD = (VD ÷ 240V) × 100

8. Continue to increase the conductor size, if necessary, until the voltage drop percentage calculates out to be 5 percent or less.
9. Enter your answers in the spaces provided on Figure 1.
10. Have your instructor check your work.

SUPPLEMENT

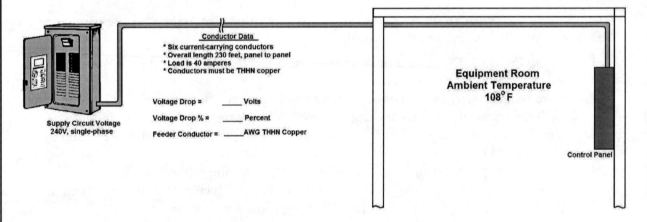

Conductor Data
* Six current-carrying conductors
* Overall length 230 feet, panel to panel
* Load is 40 amperes
* Conductors must be THHN copper

Voltage Drop = _____ Volts

Voltage Drop % = _____ Percent

Feeder Conductor = _____ AWG THHN Copper

Supply Circuit Voltage
240V, single-phase

Equipment Room
Ambient Temperature
108° F

Control Panel

Figure 1 ■ Branch Circuit Installation, Panel-to-Panel

SOLUTION

Derating the Conductor Ampacity

1. The starting point is determining what size copper THHN to use for 40 amperes when installed in an ambient temperature of 86°F and no more than 3 conductors in the conduit.

   – From *NEC Table 310.16*, we see that 10 AWG THHN copper is good for a maximum of 40 amperes.

2. Since we know that there will be derating involved based on the criteria given, we should immediately go to the next larger size, which is 8 AWG and see what its derated ampacity calculates out to be.

   – From *NEC Table 310.16*, we see that 8 AWG THHN copper is good for 55 amperes maximum.

3. Since the ambient temperature in the equipment room is estimated to be 108°F (42°C), we find the derating correction factor at the bottom of *NEC Table 310.16* to be 0.87 (87 percent) of the listed ampacity of 10 AWG copper THHN.

   – $0.87 \times 55 = 47.9$ amperes

4. We must now derate 47.9 amperes because there are 6 current-carrying conductors in the raceway.

5. We find the adjustment correction factor for 6 current-carrying conductors to be 80 percent (0.80), according to *NEC Table 310.15(B)(2)(a)*.

   – $47.9 \times 0.80 = 38.3$ amperes

6. This is less than 40 amperes, so we must use 6 AWG THHN copper as the starting point for our voltage drop calculations. If the voltage drop percentage calculated exceeds 5 percent when using 6 AWG copper dimensions, we must use the next size wire.

Calculating Voltage Drop

1. Use the formula

$$VD = \frac{2 \times L \times K \times I}{CM}$$

$$L = 230 \text{ ft}$$
$$K = 12.9 \text{ (constant)}$$
$$I = 40 \text{ amperes}$$
$$CM = ?$$

continued

2. We must find the circular mil dimension for 6 AWG stranded copper wire in *NEC Table 8, Conductor Properties* in order to complete the formula.

   – Circular mils is shown in the third-from-the-left column of *NEC Table 8*.
   – 6 AWG copper stranded circular mils = 26,240

3. We now calculate voltage drop using the formula:

$$VD = \frac{2 \times 230 \times 12.9 \times 40}{26,240}$$

$$VD = \frac{237,360}{26,240}$$

$$VD = 9.0 \text{ volts}$$

$$\%VD = 9 \div 240 \times 100$$

$$\%VD = 3.75 \text{ percent}$$

4. This is less than 5 percent, so our feeder conductors will be 6 AWG THHN copper.

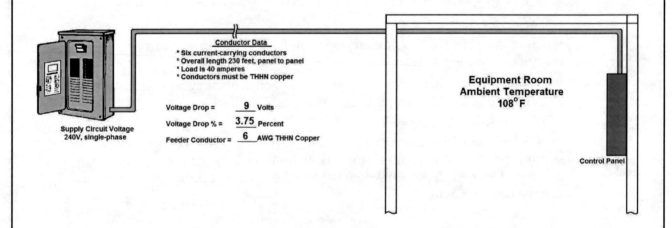

Conductor Data
* Six current-carrying conductors
* Overall length 230 feet, panel to panel
* Load is 40 amperes
* Conductors must be THHN copper

Supply Circuit Voltage
240V, single-phase

Voltage Drop = ___9___ Volts

Voltage Drop % = __3.75__ Percent

Feeder Conductor = ___6___ AWG THHN Copper

Equipment Room
Ambient Temperature
108° F

Control Panel

Module 26302-05

Project 2-2: Measure Actual Resistances of Lengths of Wire and Compare Measurements to Calculated Values Based on Resistance Values in *NEC Table 8*

Name_____ Date _____

PROJECT OVERVIEW

The *National Electrical Code®* contains tables in the back of the book that list various properties associated with conductors and conduit types. One such table is *NEC Table 8, Conductor Properties*. In this table, conductors are listed by AWG size, disregarding the type insulation such as THHN, THWN and so on, because the type of insulation does not affect the conductive capabilities of the conductor. One characteristic that does affect the conductive capabilities of a conductor is the resistance of the conductor. As you previously learned, a conductor's resistance is a major factor in voltage drop, especially in lengthy conductor installations. Voltage drop calculations are done using either a conductor's circular mil dimension or its resistance value based on one thousand feet.

The resistance columns of *NEC Table 8* include resistances for uncoated and coated copper conductors and aluminum conductors. The terms uncoated and coated do not refer to whether a conductor is insulated or not, as the information in these tables refers to bare conductors. Coated conductors are special conductors that are coated with an enamel-like substance and used in equipment and devices such as motor, transformer, and coil windings. Typical copper circuit conductors are considered as uncoated conductors.

The fourth column heading from the left in *NEC Table 8* is labeled *Quantity*. The number in this column refers to the total number of strands in the conductor. If the number is 1, the conductor is solid. Any number other than 1 refers to stranded conductors. When using this table, verify that you are referencing the correct row, as some of the smaller conductor sizes are available and listed in solid or stranded forms. Conductors larger than 8 AWG are only available in stranded form.

NEC Table 8 resistance value columns include ohms per one thousand kilometers (km) and ohms per one thousand feet (kFT). A common voltage drop formula that uses conductor resistance instead of a K-constant value is VD = 2×L×R×I ÷ 1,000. In this formula, R replaces the K-value, which is the value found in *NEC Table 8* for a known conductor size. The product of 2×L×R×I is divided by 1,000 because R is based on one thousand feet of conductor.

In this project, the trainee will measure known lengths and sizes of conductors with an accurate digital ohmmeter and will compare the measurements to calculated values.

OBJECTIVES

This performance project supports the following learning objectives listed in Module 26302-05:

- Calculate voltage drop in both single-phase and three-phase applications (*Objective 2*).
- Derate conductors for fill, temperature, and voltage drop (*Objective 6*).

Module 26302-05
Project 2-2: Measure Actual Resistances of Lengths of Wire
and Compare Measurements to Calculated Values Based on
Resistance Values in *NEC Table 8*

ELECTRICAL
PERFORMANCE PROJECT

PERFORMANCE TASKS

■ There are no performance tasks associated with this module.

MATERIALS REQUIRED

■ 1 spool (500 ft) of copper 14 AWG stranded conductor (any type)
■ 1 spool (500 ft) of copper 12 AWG stranded conductor (any type)
■ 50 ft of copper 6 AWG solid bare conductor

TOOLS AND EQUIPMENT REQUIRED

■ Pencils and paper
■ One accurate digital ohmmeter (decimal value range)
■ Calculator
■ Straightedge
■ Latest edition of the *National Electrical Code®*

REFERENCE MATERIALS AND LEARNING RESOURCES

■ *NEC Table 8, Conductor Properties*
■ Module 26302-05, Sections 1.0.0, 3.0.0, 4.1.0, 4.2.0, and 4.3.0

NOTES TO TRAINEE

■ No safety equipment is required for this project unless the environment in which the project is completed requires safety equipment.
■ Carry numbers to three decimal places.
■ Do not unspool the wire from the reels. Both ends of the conductor are usually accessible without unspooling 500 feet of wire.
■ Use a straightedge when referencing values in *NEC Table 8* to make sure you are on the right row.
■ Verify whether the conductor you are measuring is stranded or solid and reference the correct row in *NEC Table 8*.
■ Read the resistance in ohms per 1,000 feet for uncoated copper wire only.
■ Keep in mind that the resistance values in *NEC Table 8* are for 1,000 feet. Since all of the conductor lengths you are measuring are less than 1,000 feet, use the formula R = (L ÷ 1,000)(ohmskFT) to arrive at the resistance value for the length of conductor you are measuring.
■ Use Figure 1 as your answer sheet.

Module 26302-05

Project 2-2: Measure Actual Resistances of Lengths of Wire and Compare Measurements to Calculated Values Based on Resistance Values in *NEC Table 8*

NOTES TO INSTRUCTOR

- You may substitute different conductor lengths and sizes based on availability, as long as you perform the measurements and calculations prior to releasing the project to your trainees.
- The measured resistance values are very small and require an accurate digital ohmmeter that is capable of measuring in decimal values of one ohm.
- Expect measured values to differ from *NEC Table 8* calculated values. Explain to your trainees that variables such as ambient temperature, meter accuracy, and other unknown factors directly affect actual conductor resistance. The values seen in *NEC Table 8* were found using a controlled environment and a standardized measuring device. The measurements the trainees are taking represent the true resistance of the conductor for that length in that environment.
- Only the solution for calculated values for this project, based on the size and lengths of conductors listed in the material list, can be found at the end of this project.

PROCEDURE

This performance project requires you to measure the resistance of various sizes and lengths of copper conductor and compare their results with calculated resistance values according to *NEC Table 8*.

1. Use an accurate digital ohmmeter to measure total resistance from end to end for the following:
    - 500-feet 14 AWG copper conductor
    - 500-feet 12 AWG copper conductor
    - 50-feet 6 AWG copper conductor
2. Document your results in the proper locations on the answer sheet (Figure 1).
3. Refer to *NEC Table 8* and calculate the resistance in ohms per 1,000 feet, using the formula shown in your Trainee Notes for the following:
    - 500-feet 14 AWG copper conductor
    - 500-feet 12 AWG copper conductor
    - 50-feet 6 AWG copper conductor
4. Document your results in the proper locations on the answer sheet (Figure 1).
5. Calculate the difference between each of the two readings and document your results in the proper locations on the answer sheet (Figure 1).
6. Have your instructor check your work.

Module 26302-05

Project 2-2: Measure Actual Resistances of Lengths of Wire
and Compare Measurements to Calculated Values Based on
Resistance Values in *NEC Table 8*

SUPPLEMENT

Conductor	Measured	Calculated	Difference
500-ft 14 AWG			
500-ft 12 AWG			
50-ft 6 AWG			

Figure 1 ■ Answer Sheet

Module 26302-05
Project 2-2: Measure Actual Resistances of Lengths of Wire
and Compare Measurements to Calculated Values Based on
Resistance Values in *NEC Table 8*

SOLUTION

Conductor	Measured	Calculated	Difference
500-ft 14 AWG		1.57 ohms	
500-ft 12 AWG		0.99 ohms	
50-ft 6 AWG		0.025 ohms	

R = (L ÷ 1,000) × (ohmskFT)

500-ft 14 AWG:
R = (500 ÷ 1,000) x 3.14
R = 1.57 ohms

500-ft 12 AWG:
R = (500 ÷ 1,000) x 1.98
R = 0.99 ohms

50-ft 6 AWG:
R = (50 ÷ 1,000) x 0.491
R = 0.025 ohms

NOTES

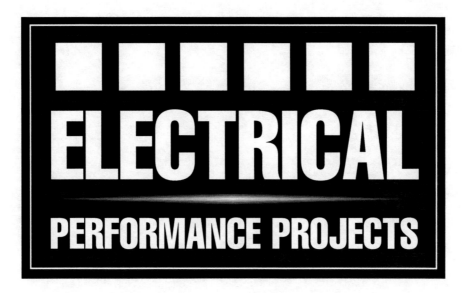

MODULE 26303-05

Projects:

- ■ **3-1** Remove and Install Fuses in a Fusible Disconnect Switch Using a Rated Fuse Puller

- ■ **3-2** Practice Selecting Fuses to Maintain Selective Coordination for a Motor Branch Circuit

Name_____ Date _____

PROJECT OVERVIEW

The primary purpose of overcurrent devices is to protect the circuit wiring in which the device is installed. All circuit wiring is rated according to the maximum current flow it can safely handle without jeopardizing the integrity of the wiring or igniting a fire. A secondary function of some overcurrent protective devices is to protect equipment in the circuit.

The two classifications of overcurrent devices are circuit breakers and fuses. The primary difference between the two is that circuit breakers can be reset to an operational condition once they have been tripped by an overcurrent condition. Fuses must be replaced.

There are many types of fuses. Some fuses are designed to instantaneously open the circuit when exposed to an overcurrent condition that exceeds their current rating. Other fuses are designed to tolerate an overcurrent condition for a short time period before opening the circuit. These are referred to as time-delay fuses. When fuses open a circuit due to an overcurrent condition, an internal conductive metal link that connects one end of the fuse to the other melts, causing the fuse to lose its continuity from one end to the other. This opens the circuit connected to the fuse.

Fuses should only be replaced and installed into existing circuits while the circuit is in a de-energized condition or safe work condition, and by using a rated fuse pulling or installing tool. Fusible disconnecting switches should be locked and tagged according to electrical safety standards.

This project requires the trainee to put a fusible disconnect in a safe work condition by locking and tagging the disconnect handle, remove the fuses using a rated fuse-pulling tool, check the fuses with a continuity meter, and replace the fuses using the rated tool.

OBJECTIVES

This performance project supports the following learning objectives listed in Module 26303-05:

- Apply the key *National Electrical Code®* requirements regarding overcurrent protection (*Objective 1*).

PERFORMANCE TASKS

- There are no performance tasks associated with this module.

MATERIALS REQUIRED

- Project board (plywood sheet or equivalent)
- One 240V or 460V fusible disconnect with fuses installed
- Sheet metal or wood screws to mount disconnect switch to board

TOOLS AND EQUIPMENT REQUIRED

- One voltage-rated fuse puller (600V)
- Approved device lock and tag
- Continuity meter or tester
- Screwdriver set or screw gun

REFERENCE MATERIALS AND LEARNING RESOURCES

- Module 26303-05, Sections 2.0.0, 2.1.0, 2.2.0, 2.3.0, 3.0.0, and 3.2.0

NOTES TO TRAINEE

- Always wear safety glasses with side shields (or goggles) when removing or installing fuses.
- Even though the fusible disconnecting switch used in this project will not have an energized circuit connected to it, you must treat it as a potentially energized switch by locking the switch handle in the OFF position and installing a tag according to your instructor's guidance.
- Do not remove or install fuses using any other tool except a rated fuse pulling/installing tool.
- When installing fuses in a single device or fuse holder, follow these guidelines:
  - Fuses must be identical, from the same manufacturer, and with the same ratings.
  - Fuses should be mounted right side up, with the identifying label facing forward.
  - Fuse holders must not be modified to make a fuse fit.

NOTES TO INSTRUCTOR

- Performing this project on existing, installed fused disconnect switches or devices is not recommended.
- Demonstrate proper lockout/tagout procedures to your trainees if they have not been exposed to these procedures.
- Observe and evaluate trainees during the fuse removal and installation process, looking for accidental hand contact with simulated energized parts.
- Verify that continuity checks are made by the trainees on all fuses once the fuses have been removed.
- Observe and evaluate the positions of the fuses after installation.

PROCEDURE

This performance project requires you to put a fusible disconnecting switch in a safe work condition, remove fuses, check fuse continuity, and install the fuses.

1. Refer to Figure 1.
2. Mount the disconnecting switch to the project board.
3. Observe proper lockout/tagout procedures presented by your instructor.
4. Properly lock and tag the fusible disconnecting switch handle in the OFF position
5. Use the fuse puller to remove the fuses from the disconnecting switch.
6. Use a continuity tester or ohmmeter to test continuity on all fuses.
7. Reinstall the fuses in the disconnecting switch using the fuse tool.
8. Have your instructor check your work.

SUPPLEMENT

Figure 1 ■ Using a Rated Fuse Puller

Name_____ Date _____

PROJECT OVERVIEW

At startup, a motor can draw more than five times the amount of its normal operating current. Because of this factor, dual-element time-delay fuses are typically used in branch motor circuits to protect motors against overcurrent. Dual-element time-delay fuses may be more closely sized to the full load running amperage of the motor than nontime-delay fuses, because the time-delay fuses can withstand the short duration of increased starting current without blowing. These fuses also provide good overcurrent protection for the normal running cycles. In contrast, nontime-delay fuses may be sized more than five times the running amperage to prevent them from blowing during motor startup. Such fuses provide very little protection from overcurrent during normal running cycles.

Fuses used as overcurrent protection in feeder circuits that supply branch motor circuits must be rated larger than the branch circuit fuses. This prevents both fuses from blowing should an overcurrent condition take place in the branch motor circuit. The ratio between upstream fuse ratings and downstream fuse ratings is referred to as selective coordination. Ideally, if a nontime-delay fuse is installed in the motor branch circuit, a ratio of 3:1 should be maintained between the feeder and branch circuit fuse ratings. However, if dual-element time-delay fuses are installed in both the feeder circuit and branch circuit, the ratio can be reduced to 2:1.

In this project the trainee will refer to a one-line drawing of a feeder circuit and a motor branch circuit containing fuses. The trainee must determine the minimum size dual-element time-delay fuse that can be installed in the feeder circuit. The trainee must also answer three questions associated with the circuit and motor circuit protection.

OBJECTIVES

This performance project supports the following learning objectives listed in Module 26303-05:

- Apply the key *National Electrical Code®* requirements regarding overcurrent protection (*Objective 1*).
- Check specific applications for conformance to *NEC®* sections that cover short circuit current, fault currents, interrupting ratings, and other sections relating to overcurrent protection (*Objective 2*).
- Select and size overcurrent protection for specific applications (*Objective 4*).

PERFORMANCE TASKS

- There are no performance tasks associated with this module.

MATERIALS REQUIRED

- None

ELECTRICAL
PERFORMANCE PROJECT

TOOLS AND EQUIPMENT REQUIRED

- Pencils and paper
- Calculator

REFERENCE MATERIALS AND LEARNING RESOURCES

- Module 26303-05, Sections 1.1.0, 1.1.1, 2.0.0, 2.1.0, 2.5.0, 3.1.0, 3.2.0, 4.4.0, and 5.0.0.;
  Table 3, Selection of fuses for motor protection (2 of 3).
- *NEC Sections 240.6 and 430.72*

NOTES TO TRAINEE

- No safety equipment is required for this project unless the environment in which the
  project is completed requires safety equipment.
- You must select the minimum value fuse rating that corresponds to a standard fuse
  ampere rating based on the recommended selective coordination ratio. Standard fuse
  ampere ratings may be found in the *NEC Section 240.6*.
- Answer the three questions associated with Figure 1 and motor overload protection.

NOTES TO INSTRUCTOR

- The solutions can be found at the end of this project.

PROCEDURE

This performance project requires you to review a one-line feeder and motor branch circuit
drawing and determine the minimum value fuse rating for the feeder circuit based on the
recommended selective coordination ratio. The trainee must also answer three associated
questions by referring to Figure 1 and tables in Module 26303-05.

1. Review Figure 1, noting the size and type of fuses in the feeder and motor branch
   circuits.
2. Review your text if necessary to determine the recommended selective coordination
   ratio between upstream and downstream overcurrent protective fuses.
3. Determine the minimum value fuse rating for Fuse A that corresponds to a standard
   fuse ampere rating, based on the recommended ratio and the type of fuse in the motor
   branch circuit (Fuse B). Standard fuse ampere ratings may be found in the *NEC Section
   240.6*.
4. Answer the three questions associated with Figure 1 and motor overcurrent protection.
5. Have your instructor check your work.

SUPPLEMENT

480V, THREE-PHASE

FUSE A

TIME-DELAY DUAL-
ELEMENT FUSE
_____ AMPS

FUSE B

TIME-DELAY DUAL-
ELEMENT FUSE
25 AMPS

460V, THREE-PHASE
15-HP MOTOR
TEMPERATURE RISE 40°C
FLA 21 AMPS

QUESTIONS

1. WHAT IS THE MINIMUM SELECTIVE COORDINATION RATIO
   ALLOWED BETWEEN FUSE A AND FUSE B IN THIS INSTALLATION?

   _____ WHY?_____

2. BASED ON THE VALUE OF FUSE B AND TABLE 3 LOCATED IN
   MODULE 26303-05, IS THIS MOTOR PROVIDED WITH BACKUP
   PROTECTION IN THE FORM OF PROPERLY SIZED OVERLOAD
   RELAYS?

   _____ YES   _____ NO

3. BASED ON THE VALUE OF FUSE B AND TABLE 3 LOCATED IN
   MODULE 26303-05, IS THE SERVICE FACTOR OF THIS MOTOR
   LESS THAN OR GREATER THAN 1.15?

   _____ LESS THAN   _____ GREATER THAN

Figure 1 ■ One-Line Motor Circuit and Related Questions

SOLUTION

480V, THREE-PHASE

FUSE A

TIME-DELAY DUAL-
ELEMENT FUSE
_____ AMPS

FUSE B

TIME-DELAY DUAL-
ELEMENT FUSE
25 AMPS

460V, THREE-PHASE
15-HP MOTOR
TEMPERATURE RISE 40°C
FLA 21 AMPS

QUESTIONS

1. WHAT IS THE MINIMUM SELECTIVE COORDINATION RATIO
   ALLOWED BETWEEN FUSE A AND FUSE B IN THIS INSTALLATION?

   _____2:1_____ WHY?_____*Fuses are time delay*_____

2. BASED ON THE VALUE OF FUSE B AND TABLE 3 LOCATED IN
   MODULE 26303-05, IS THIS MOTOR PROVIDED WITH BACKUP
   PROTECTION IN THE FORM OF PROPERLY SIZED OVERLOAD
   RELAYS?

   _____ YES _____X_____ NO

3. BASED ON THE VALUE OF FUSE B AND TABLE 3 LOCATED IN
   MODULE 26303-05, IS THE SERVICE FACTOR OF THIS MOTOR
   LESS THAN OR GREATER THAN 1.15?

   _____ LESS THAN _____X_____ GREATER THAN

ANSWER KEY

1. THE RATIO IS 2:1 BECAUSE FUSE B IS A DUAL-ELEMENT TIME-DELAY FUSE. IF FUSE B WERE A NONTIME-DELAY
   FUSE, THE RATIO WOULD HAVE TO BE 3:1.

2. THE ANSWER IS FOUND IN TABLE 3 (2 OF 3) LOCATED IN MODULE 26303-05. FUSE B IS RATED AT 25 AMPERES.
   THIS MATCHES THE FIRST COLUMN FOR MOTORS HAVING A TEMPERATURE RISE OF NOT OVER 40°C WITHOUT
   PROPERLY SIZED OVERLOAD RELAYS.

3. THE SAME AS QUESTION 2, BUT RELATED TO SERVICE FACTORS 1.15 OR GREATER.

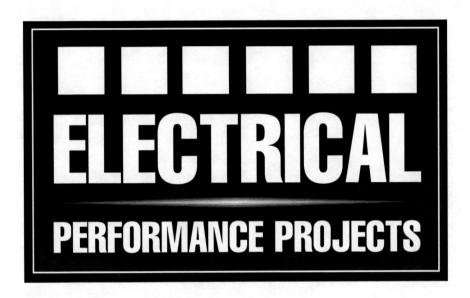

MODULE 26304-05

Projects:

- **4-1 Calculate the Minimum Conduit and Box Size Requirements Based on a Set of Conditions**

- **4-2 Size a Pull Box Based on the Number and Size of Conduits Entering the Box**

Name_____ Date _____

PROJECT OVERVIEW

Conduit fill requirements are based on the areas of a conduit and the sum total of the areas of all conductors that are to be installed in the conduit. Think of conduit area as the circle that is formed by the interior walls of the conduit and conductor area as the outer circle that is formed by the insulation around a conductor. Unlike the resistive properties of the conductor, the type of conductor insulation does directly affect conductor area. According to the *National Electrical Code®*, only 40 percent of the conduit's interior area may be filled by conductors when three or more conductors are installed in the conduit.

Tables in the back of the *NEC®* list conductor fill properties of conduit according to trade and also list the areas of conductors based on size and type of conductor, such as THHN, THWN, and so on. The *NEC®* also contains tables that list the maximum number of a specific conductor size and type that is allowed in a specific trade size conduit. However, if more than one size or type of conductor is to be installed in the same conduit, conduit fill must be calculated by first calculating the sum total of the areas of all the conductors that are to be installed in the conduit.

The conductor fill of boxes installed in conduit systems is also regulated by the *NEC®*. One major difference between conduit fill and box fill is that conduit fill is based on area or inches squared, while box fill is based on volume or inches cubed. Box fill tables are not located in the back of the *NEC®*, but are located in *NEC Article 314. NEC Table 314.16(A)* lists common box sizes, their total volume capacity, and the maximum number of a specific AWG size conductor that may be installed in each box. *NEC Table 314.16(B)* lists the volume allowances required for conductors based on their AWG size. If more than one conductor size is to be installed in a single box, box fill must be calculated by first calculating the sum total of the volume allowance requirements for all conductors to be installed. The trainee will calculate conduit and box fill in this project.

OBJECTIVES

This performance project supports the following learning objectives listed in Module 26304-05:

- Size raceways according to conductor fill and *NEC®* installation requirements (*Objective 1*).
- Size outlet boxes according to *NEC®* installation requirements (*Objective 2*).
- Calculate conduit fill using a percentage of the trade size conduit inside diameter (ID) (*Objective 3*).

PERFORMANCE TASKS

- There are no performance tasks associated with this module.

Module 26304-05
Project 4-1: Calculate the Minimum Conduit and
Box Size Requirements Based on a Set of Conditions

ELECTRICAL
PERFORMANCE PROJECT

MATERIALS REQUIRED

- None

TOOLS AND EQUIPMENT REQUIRED

- Pencils and paper
- Calculator
- Latest edition of the *National Electrical Code®*

REFERENCE MATERIALS AND LEARNING RESOURCES

- *NEC Section 314.16(A); Tables 314.16(A) and (B); Chapter 9, Tables 4 and 5*
- Module 26304-05, Sections 2.0.0, 2.1.0, 4.0.0, and 4.1.0

NOTES TO TRAINEE

- No safety equipment is required for this project unless the environment in which the project is completed requires safety equipment.
- You must reference the appropriate sections and tables in the *NEC®* in order to complete this project.

NOTES TO INSTRUCTOR

- Trainees must use the *NEC®* to complete this project.
- All conduit runs shown are assumed to be 10 feet or longer and are not considered as conduit nipples.
- Evaluate the trainees's work based on the minimum sizes only. Do not allow the trainees to oversize conduit or box sizes because there is no way to evaluate their calculations when this is done.
- The solutions can be found at the end of this project.

PROCEDURE

This performance project requires you to calculate conduit and box fill based on information located on Figure 1.

1. Calculate the minimum size electrical metallic tubing (EMT) that must be installed based on the number, size, and type of conductors shown in Figure 1.
2. Calculate the minimum size square boxes that must be installed in each location based on the number and size of conductors entering and leaving the boxes from both conduit and cables.
3. Write your answers in the spaces provided in Figure 1.

Module 26304-05
Project 4-1: Calculate the Minimum Conduit and
Box Size Requirements Based on a Set of Conditions

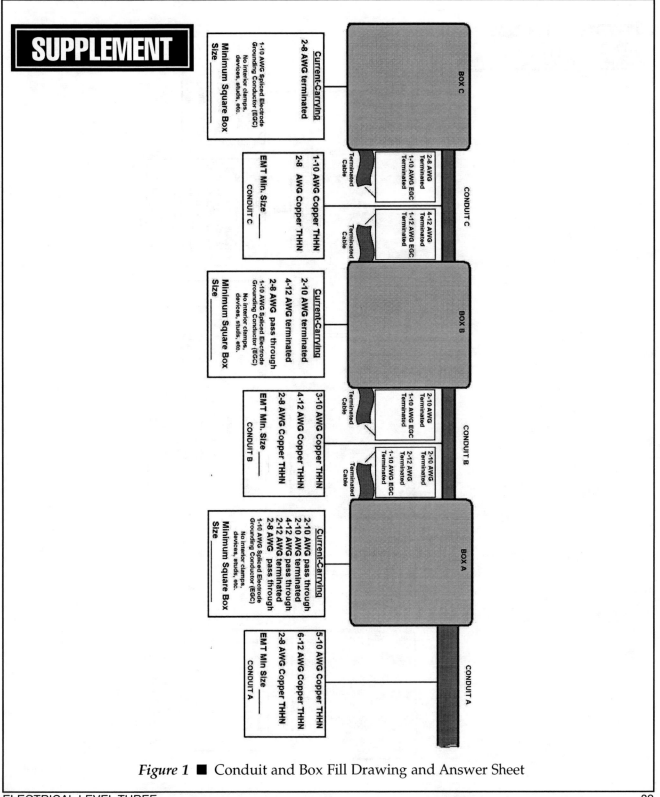

Figure 1 ■ Conduit and Box Fill Drawing and Answer Sheet

Module 26304-05
Project 4-1: Calculate the Minimum Conduit and
Box Size Requirements Based on a Set of Conditions

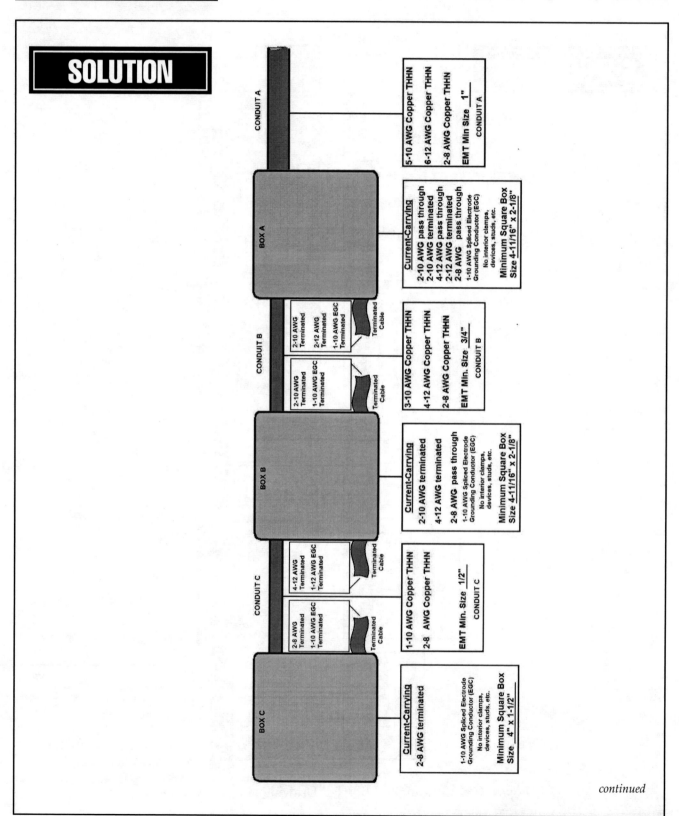

SOLUTION

CONDUIT A

5-10 AWG Copper THHN
6-12 AWG Copper THHN
2-8 AWG Copper THHN
EMT Min Size __1"__
CONDUIT A

BOX A

Current-Carrying
2-10 AWG pass through
2-10 AWG terminated
4-12 AWG pass through
2-12 AWG terminated
2-8 AWG pass through
1-10 AWG Spliced Electrode
Grounding Conductor (EGC)
No interior clamps,
devices, studs, etc.
Minimum Square Box
Size 4-11/16" x 2-1/8"

CONDUIT B

2-10 AWG Terminated
2-12 AWG Terminated
1-10 AWG EGC Terminated
Terminated Cable

2-10 AWG Terminated
1-10 AWG EGC Terminated
Terminated Cable

3-10 AWG Copper THHN
4-12 AWG Copper THHN
2-8 AWG Copper THHN
EMT Min. Size __3/4"__
CONDUIT B

BOX B

Current-Carrying
2-10 AWG terminated
4-12 AWG terminated
2-8 AWG pass through
1-10 AWG Spliced Electrode
Grounding Conductor (EGC)
No interior clamps,
devices, studs, etc.
Minimum Square Box
Size 4-11/16" x 2-1/8"

CONDUIT C

4-12 AWG Terminated
1-12 AWG EGC Terminated
Terminated Cable

2-8 AWG Terminated
1-10 AWG EGC Terminated
Terminated Cable

1-10 AWG Copper THHN
2-8 AWG Copper THHN
EMT Min. Size __1/2"__
CONDUIT C

BOX C

Current-Carrying
2-8 AWG terminated
1-10 AWG Spliced Electrode
Grounding Conductor (EGC)
No interior clamps,
devices, studs, etc.
Minimum Square Box
Size __4" x 1-1/2"__

continued

Conduit Sizing by Area

Conduit A:

5 – 10 AWG THHN at 0.0211 in² each	= 0.1055
6 – 12 AWG THHN at 0.0133 in² each	= 0.0798
2 – 8 AWG THHN at 0.0366 in² each	= 0.0732
Total in²	= 0.2585 in²

NEC Chapter 9 Table 4, Dimensions and Percent Area of Conduit and Tubing (Areas of Conduit or Tubing for the Combinations of Wires Permitted in Table 1, Chapter 9).

¾" EMT conduit – over 2 wires (40 percent)	= 0.213 in² (too small)
1" EMT conduit – over 2 wires (40 percent)	**= 0.346 in² (large enough)**

Conduit B:

3 – 10 AWG THHN at 0.0211 in² each	= 0.0633
4 – 12 AWG THHN at 0.0133 in² each	= 0.0532
2 – 8 AWG THHN at 0.0366 in² each	= 0.0732
Total in²	= 0.1897 in²

NEC Chapter 9 Table 4: Dimensions and Percent Area of Conduit and Tubing (Areas of Conduit or Tubing for the Combinations of Wires Permitted in Table 1, Chapter 9).

½" EMT conduit – over 2 Wires (40 percent)	= 0.122 in² (too small)
¾" EMT conduit – over 2 Wires (40 percent)	**= 0.213 in² (large enough)**

Conduit C:

1 – 10 AWG THHN at 0.0211 in² each	= 0.0211
2 – 8 AWG THHN at 0.0366 in² each	= 0.0732
Total in²	= 0.0943 in²

continued

Module 26304-05
Project 4-1: Calculate the Minimum Conduit and
Box Size Requirements Based on a Set of Conditions

Box Sizing by Volume

Box A:

NEC Table 314.16(B), Conductors Entering and Leaving from Conduits

2 – 10 AWG THHN pass thru at 2.50 in³ each	= 5.00
2 – 10 AWG THHN terminated at 2.50 in³ each	= 5.00
4 – 12 AWG THHN pass thru at 2.25 in³ each	= 9.00
2 – 12 AWG THHN terminated at 2.25 in³ each	= 4.50
2 – 8 AWG THHN pass thru at 3.00 in³ each	= 6.00

Conductors Entering and Leaving from Cables

2 – 10 AWG THHN terminated at 2.50 in³ each	= 5.00
2 – 12 AWG THHN terminated at 2.25 in³ each	= 4.50

Electrode Grounding Conductors (EGCs are only counted as one)

1 – 10 AWG THHN spliced at 2.50 in³ each	= 2.50
Total in³	= 41.5 in³

NEC Article 314, Table 314.16(A), Metal Boxes

4¹¹⁄₁₆" × 1½" square	= 29.5 in³ (too small)
4¹¹⁄₁₆" × 2⅛" square	**= 42.0 in³ (large enough)**

Box B:

Conductors Entering and Leaving from Conduits

2 – 10 AWG THHN terminated at 2.50 in³ each	= 5.00
4 – 12 AWG THHN terminated at 2.25 in³ each	= 9.00
2 – 8 AWG THHN pass thru at 3.00 in³ each	= 6.00

Conductors Entering and Leaving from Cables

2 – 10 AWG THHN terminated at 2.50 in³ each	= 5.00
4 – 12 AWG THHN terminated at 2.25 in³ each	= 9.00

Electrode Grounding Conductors (EGCs are only counted as one)

1 – 10 AWG THHN Spliced @ 2.50 in³ each	= 2.50
Total in³	= 36.5 in³

NEC Article 314, Table 314.16(A), Metal Boxes

4¹¹⁄₁₆" × 1½" square	= 29.5 in³ (too small)
4¹¹⁄₁₆" × 2⅛" square	**= 42.0 in³ (large enough)**

continued

Module 26304-05
Project 4-1: Calculate the Minimum Conduit and
Box Size Requirements Based on a Set of Conditions

Box C:

Conductors Entering and Leaving from Conduits

2 – 8 AWG THHN terminated at 3.00 in^3 each   = 6.00

Conductors Entering and Leaving from Cables

2 – 8 AWG THHN terminated at 3.00 in^3 each   = 6.00

Electrode Grounding Conductors (EGCs are only counted as one)

1 – 10 AWG THHN spliced at 2.50 in^3 each   = 2.50
Total in^3 =14.50 in^3

NEC Article 314, Table 314.16(A), Metal Boxes

4" × 1¼" square   = 12.5 in^3 (too small)
4" x 1½" square   **= 15.5 in^3 (large enough)**

ELECTRICAL
PERFORMANCE PROJECT

NOTES

Module 26304-05
**Project 4-2: Size a Pull Box Based on the Number
and Size of Conduits Entering the Box**

Name_____ **Date** _____

PROJECT OVERVIEW

The *National Electrical Code®* regulates the size of boxes that are used as pull or junction boxes in conduit installations. If the conductors to be installed in conduits entering and leaving a junction or pull box are 4 AWG or larger, *NEC®* regulations are in place to cover box sizes associated with both straight and angle pulls.

In straight pulls the length of the box from one wall to the other must be at least eight times the trade size of the largest conduit.

In angle pulls different rules apply. These rules can be found in *NEC Section 314.28(A)(2)*.

In this project, the trainee will calculate box size based on an angle pull.

OBJECTIVES

This performance project supports the following learning objectives listed in Module 26304-05:

- Size and select pull and junction boxes according to *NEC®* installation requirements (*Objective 3*).
- Calculate the required bending radius in boxes and cabinets (*Objective 4*).

PERFORMANCE TASKS

- There are no performance tasks associated with this module.

MATERIALS REQUIRED

- None

TOOLS AND EQUIPMENT REQUIRED

- Pencils and paper
- Calculator
- Latest edition of the *National Electrical Code®*

REFERENCE MATERIALS AND LEARNING RESOURCES

- *NEC Section 314.28*
- Module 26304-05, Section 3.1.0

Module 26304-05
**Project 4-2: Size a Pull Box Based on the Number
and Size of Conduits Entering the Box**

NOTES TO TRAINEE

■ No safety equipment is required for this project unless the environment in which the project is completed requires safety equipment.
■ All conductors to be installed in the conduits shown are assumed to be 4 AWG or larger.
■ Round off box sizes to whole numbers (no decimal places).
■ You should reference the appropriate section in the *NEC*® in order to complete this project.

NOTES TO INSTRUCTOR

■ All conductors to be installed in the conduits shown are assumed to be 4 AWG or larger.
■ The solution is provided at the end of this project.

PROCEDURE

This performance project requires you to calculate a pull box size in which the conductors are pulled at an angle, based on the number and size of conduits entering and leaving the box as shown in Figure 1.

1. Calculate L1 by multiplying the largest conduit size by six and adding to the results the sum of the other conduit sizes entering the box on the same side.
2. Calculate L2 in the same manner as L1.
3. Calculate distance (D) between raceway entries enclosing the same conductors by multiplying the largest conduit size by six.
4. Show your answers directly on Figure 1.
5. Have your instructor check your work.

SUPPLEMENT

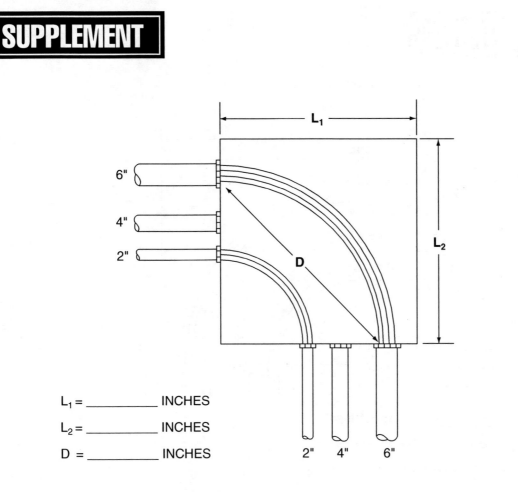

L₁ = _____ INCHES

L₂ = _____ INCHES

D = _____ INCHES

Figure 1 ■ Pull Box Drawing

Module 26304-05
**Project 4-2: Size a Pull Box Based on the Number
and Size of Conduits Entering the Box**

SOLUTION

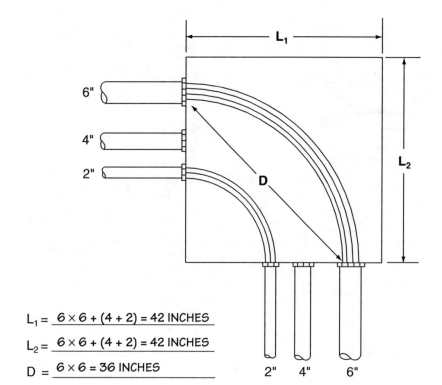

$L_1 = \underline{\;6 \times 6 + (4 + 2) = 42\ \text{INCHES}\;}$

$L_2 = \underline{\;6 \times 6 + (4 + 2) = 42\ \text{INCHES}\;}$

$D = \underline{\;6 \times 6 = 36\ \text{INCHES}\;}$

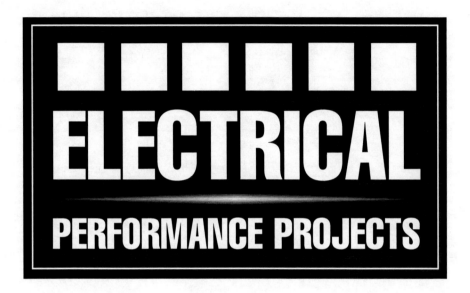

ELECTRICAL
PERFORMANCE PROJECTS

MODULE 26305-05

Projects:

- ■ **5-1** Install Device Boxes and Receptacles on a Simulated Residential Countertop and Wire the Receptacles to Provide GFCI Protection from a Single GFCI Receptacle

- ■ **5-2** Identifying and Locating Receptacles from a Residential Floor Plan

ELECTRICAL

PERFORMANCE PROJECT

Module 26305-05

Project 5-1: Install Device Boxes and Receptacles on a Simulated Residential Countertop and Wire the Receptacles to Provide GFCI Protection from a Single GFCI Receptacle

Name_____ Date _____

PROJECT OVERVIEW

The *National Electrical Code®* generally does not regulate mounting heights for receptacles in residential locations. An exception is the requirement that receptacles above kitchen countertops must not be installed more than twenty inches above the countertop or more than twelve inches below the countertop surface where this condition is possible. These height regulations are found in *NEC Section 210.52*. The *NEC®* also states that receptacles installed to serve a residential kitchen countertop must be placed so that no point along the wall line is more than twenty-four inches, measured horizontally, from a receptacle outlet in that countertop space.

The *NEC®* requires all receptacles that serve the countertop in residential dwellings be equipped with ground fault protection. There are two common practices used to comply with this regulation: installing GFCI-type receptacles in each device box serving a kitchen countertop and installing one GFCI-type receptacle, then installing standard duplex receptacles in the rest of the countertop device boxes, providing ground-fault protection through the GFCI-type receptacle.

In this project, the trainee will create a simulated kitchen countertop installation, install device boxes and receptacles according to *NEC®* regulations, and provide ground-fault protection for all receptacles through one GFCI-type receptacle.

OBJECTIVES

This performance project supports the following learning objectives listed in Module 26305-05:

■ Follow *NEC®* regulations governing the installation of wiring devices (*Objective 4*).

PERFORMANCE TASKS

This performance project supports the following performance task(s) listed for Module 26305-05:

■ Identify different types of receptacles (*Task 1*).
■ Select the appropriate receptacle for a given application (*Task 2*).

MATERIALS REQUIRED

■ One project board (sheet of plywood, sheetrock, or equivalent)
■ Four plastic single-gang pop-in device boxes
■ Three standard duplex receptacles
■ Three plastic or nylon duplex receptacle plates (any color)
■ One standard GFCI duplex receptacle

continued

ELECTRICAL

PERFORMANCE PROJECT

- One plastic or nylon GFCI duplex receptacle plate (any color)
- 20 to 25 feet of 12/2 NM cable with ground
- One 120-volt pigtail with cap (plug)
- One roll of duct tape

TOOLS AND EQUIPMENT REQUIRED

- Marker or pencils
- Tape measure
- 6-ft straight edge or chalk line
- Torpedo level
- Keyhole saw or jigsaw
- Lineman's pliers
- Screwdriver set
- Wire strippers

REFERENCE MATERIALS AND LEARNING RESOURCES

- *NEC Sections 210.8 and 210.52*
- Module 26305-05, Sections 2.0.0, 2.2.0, 2.3.0, 3.0.0, and 3.1.0

NOTES TO TRAINEE

- This project requires safety glasses with side shields (or goggles) to be worn at all times.
- Wear work gloves when cutting, sawing, and working with NM cable.
- When marking the project board for device box locations, place the box opening against the board, level the box with the torpedo level, and mark the board for cutout.
- Make sure that the box cutouts are sized so that the device box flanges rest against the outside surface of the board as the pop-in box wings are tightened to secure the box in place.
- Install all receptacles with the ground terminals down.
- Do not use backstab terminals (if equipped) on receptacles to connect wires. Make all receptacle connections using the terminal screws.

NOTES TO INSTRUCTOR

- Device box types may be substituted based on availability.
- This project may be enhanced by constructing a plywood and 2' × 4' frame countertop, sink cut-out, and a plywood and frame backsplash to create a more realistic kitchen scenario.

continued

Module 26305-05

Project 5-1: Install Device Boxes and Receptacles on a Simulated Residential Countertop and Wire the Receptacles to Provide GFCI Protection from a Single GFCI Receptacle

- Discourage using the backstab terminals on receptacles, but do encourage using the screw terminals for termination integrity.
- Demonstrate to the trainees the proper way to wrap the conductor ends around the terminal screw, in the direction of screw tightening.
- Once the circuit has been energized by plugging in the AC pigtail, have trainees test the GFCI receptacle by pressing the TEST button.
- After pressing the GFCI TEST button, have the trainees ascertain that the rest of the receptacles have been de-energized through the activation of the one GFCI receptacle. If any receptacles show power after the GFCI TEST button has been pressed, a wiring problem exists somewhere in the circuit. De-energize the circuit and have the trainees troubleshoot the circuit.
- Remember to press the RESET button on the GFCI receptacle to restore power to the receptacles.
- Receptacles should be mounted in a vertically level position.

PROCEDURE

This performance project requires you to install device boxes and receptacles on a simulated residential kitchen countertop according to *NEC®* regulations.

1. Refer to Figure 1 to complete the following steps.
2. Prepare the project board by placing a piece of duct tape lengthwise and level across the center of the board to simulate the countertop's edges.
3. Evenly space off a 2' × 4' rectangle to represent the kitchen sink on the bottom half of the board, keeping in mind that the *NEC®* considers the sink as a break in the horizontal wall line of the countertop.
4. Position and mark the box cutouts in compliance with *NEC®* regulations
5. Cut out the rectangles for the pop-in boxes.
6. Install the four pop-in boxes, and secure in place with the integral winged fasteners.
7. Install NM cable between the four boxes as shown.
8. Install the 120-volt pigtail in the far left box as shown.
9. Install and wire the receptacles as shown.
10. Have your instructor check your wiring work.
11. Install the cover plates.
12. Energize the installation and check your work.

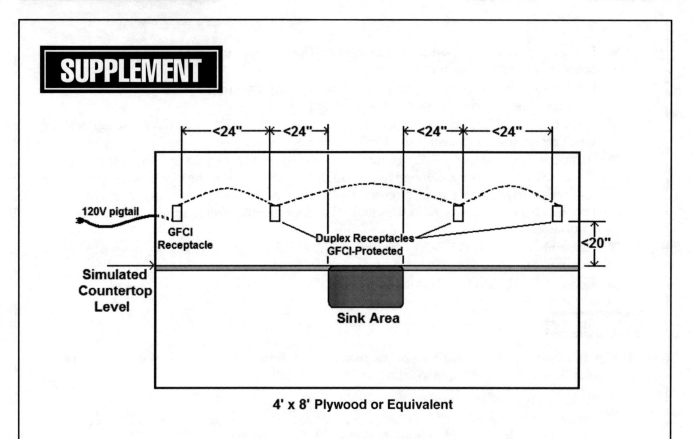

Figure 1 ■ Simulated Residential Kitchen Countertop

Name_____ Date _____

PROJECT OVERVIEW

Electrical planning and design are not always incorporated into electrical installations, especially on smaller residential jobs. Often, it is left up to the electrician's knowledge and experience in determining the type, locations, and number of receptacles that must be installed. Electricians who are fortunate enough to have electrical floor plans provided can select, place, and wire receptacles at a much faster pace than those who must do it all on their own and still remain compliant with the *NEC*® requirements.

NEC Sections 210.8 and *210.52* cover receptacle installation requirements for residences. Ground fault protection, either in the form of GFCI receptacles or branch circuit breakers, is required in certain locations in and around the house. Kitchen receptacles must comply with certain requirements as to minimum and maximum heights. The *NEC*® does not regulate the maximum number of receptacles that may be installed on a residential branch circuit, but it does establish the minimum number of small-appliance kitchen branch circuits as two 20A circuits. The *NEC*® also requires that at least one 20A branch circuit be installed in the laundry room. Other regulations address residential receptacle installations and use in bathrooms, outdoors, basements and garages, hallways, island countertops, and laundry areas.

In this project, the trainee will reference a floor plan drawing and determine the types of receptacles on given branch circuits represented by circuit numbers listed on the drawing. The trainee must also identify the locations of the receptacles, and whether the receptacle is located on a small appliance branch circuit, laundry room branch circuit, or neither.

OBJECTIVES

This performance project supports the following learning objectives listed in Module 26305-05:

■ Follow *NEC*® regulations governing the installation of wiring devices (*Objective 4*).

PERFORMANCE TASKS

This performance project supports the following performance task(s) listed for Module 26305-05:

■ Identify different types of receptacles (*Task 1*).
■ Select the appropriate receptacle for a given application (*Task 3*).

MATERIALS REQUIRED

- Floor plan (Figure 1)
- Pencil
- Highlighter

TOOLS AND EQUIPMENT REQUIRED

- Latest edition of the *National Electrical Code®*

REFERENCE MATERIALS AND LEARNING RESOURCES

- *NEC Sections 210.8* and *210.52*
- Module 26305-05, 2.0.0, 2.1.0, 2.2.0, 2.3.0, 3.0.0, and 3.1.0

NOTES TO TRAINEE

- No safety equipment is required for this project unless the environment in which the project is completed requires safety equipment.
- Arrowheads on the drawing indicate a homerun or branch circuit termination, which is indicated by a circuit number that corresponds to the branch circuit location in Panel A located in the Utility Room.
- Locate the receptacles on a branch circuit by starting at the arrowhead, following the circuit routing line and identifying each receptacle attached to that line. Do not confuse circuit routing lines with design lines.
- Not all circuits start and end in the same room.
- It may help you to identify all receptacles by highlighting the circuit routing lines and receptacles as you complete each branch circuit.
- When identifying receptacles in the kitchen or dining area, specify if the receptacle is serving the countertop.
- You must know or reference the regulations in *NEC Section 210.8* relating to GFCI requirements in order to complete this project.

NOTES TO INSTRUCTOR

- The instructor's solution is located at the end of this project.

PROCEDURE

This performance project requires you to reference a floor plan drawing and determine the type of receptacles, either standard or GFCI, on a given branch circuit represented by a circuit number on the drawing. You must also identify the location of the receptacle, and determine whether the receptacle is located on a small appliance branch circuit, laundry room branch circuit, or neither.

1. Refer to Figure 1, Electrical Floor Plan.
2. Locate each branch circuit by circuit number, starting with Ckt. 1 and ending with Ckt. 8.
3. Identify each receptacle, based on location and *NEC*® requirements, as either standard or GFCI-protected.

Note:

There are thirty-one receptacles on this floor plan.

4. Identify the room or area in which the receptacle is located.
5. Determine if the receptacle is on a small appliance branch circuit, laundry circuit, or neither.
6. Write your answers on the answer sheet provided in Figure 2.
7. Have your instructor check your work.

SUPPLEMENT

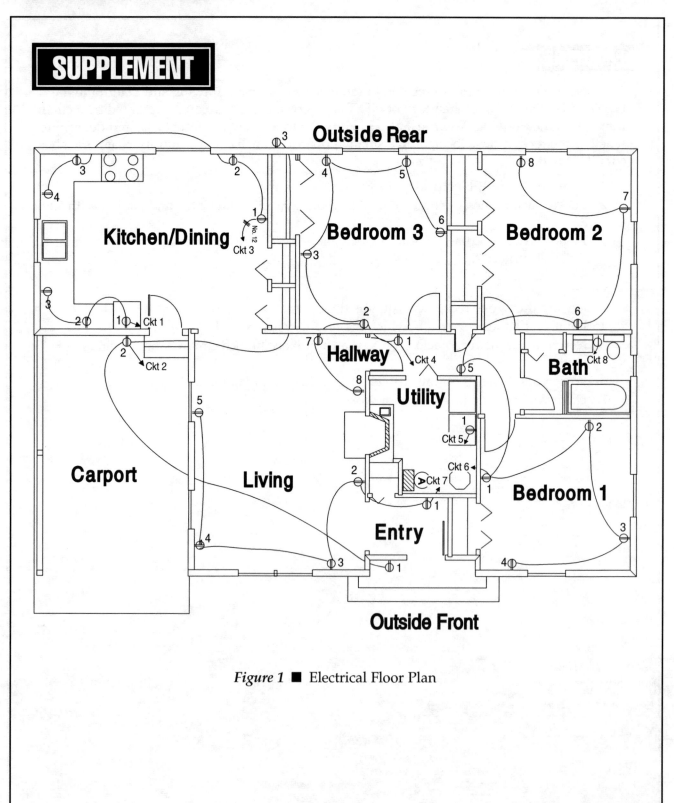

Figure 1 ■ Electrical Floor Plan

SUPPLEMENT

Circuit Number	Receptacle Number	Recept. Type Stand. or GFCI	Room Location	Small Appliance, Laundry, Neither
Ckt 1	1			
Ckt 1	2			
Ckt 1	3			
Ckt 2	1			
Ckt 2	2			
Ckt 2	3			
Ckt 3	1			
Ckt 3	2			
Ckt 3	3			
Ckt 3	4			
Ckt 4	1			
Ckt 4	2			
Ckt 4	3			
Ckt 4	4			
Ckt 4	5			
Ckt 4	6			
Ckt 5	1			
Ckt 6	1			
Ckt 6	2			
Ckt 6	3			
Ckt 6	4			
Ckt 6	5			
Ckt 6	6			
Ckt 6	7			
Ckt 6	8			
Ckt 7	1			
Ckt 7	2			
Ckt 7	3			
Ckt 7	4			
Ckt 7	5			
Ckt 8	1			

Figure 2 ■ Answer Sheet

SOLUTION

Circuit Number	Receptacle Number	Recept. Type Stand. or GFCI	Room Location	Small Appliance, Laundry, Neither
Ckt 1	1	GFCI	Kitchen/Countertop	Small Appliance
Ckt 1	2	GFCI	Kitchen/Countertop	Small Appliance
Ckt 1	3	GFCI	Kitchen/Countertop	Small Appliance
Ckt 2	1	GFCI	Outside Front	Neither
Ckt 2	2	GFCI	Carport	Neither
Ckt 2	3	GFCI	Outside Rear	Neither
Ckt 3	1	Standard	Kitchen/Dining	Small Appliance
Ckt 3	2	Standard	Kitchen/Dining	Small Appliance
Ckt 3	3	GFCI	Kitchen/Countertop	Small Appliance
Ckt 3	4	GFCI	Kitchen/Countertop	Small Appliance
Ckt 4	1	Standard	Hallway	Neither
Ckt 4	2	Standard	Bedroom 3	Neither
Ckt 4	3	Standard	Bedroom 3	Neither
Ckt 4	4	Standard	Bedroom 3	Neither
Ckt 4	5	Standard	Bedroom 3	Neither
Ckt 4	6	Standard	Bedroom 3	Neither
Ckt 5	1	Standard	Utility	Laundry
Ckt 6	1	Standard	Bedroom 1	Neither
Ckt 6	2	Standard	Bedroom 1	Neither
Ckt 6	3	Standard	Bedroom 1	Neither
Ckt 6	4	Standard	Bedroom 1	Neither
Ckt 6	5	Standard	Hallway	Neither
Ckt 6	6	Standard	Bedroom 2	Neither
Ckt 6	7	Standard	Bedroom 2	Neither
Ckt 6	8	Standard	Bedroom 2	Neither
Ckt 7	1	Standard	Entry	Neither
Ckt 7	2	Standard	Living	Neither
Ckt 7	3	Standard	Living	Neither
Ckt 7	4	Standard	Living	Neither
Ckt 7	5	Standard	Living	Neither
Ckt 8	1	GFCI	Bath	Neither

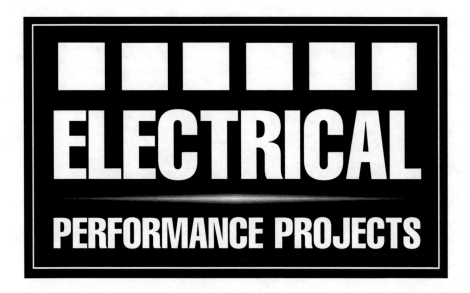

MODULE 26306-05

Projects:

PROJECT OVERVIEW

There are several major spacing concerns that must be considered when installing switchgear and its internal parts. These considerations include work spaces around the outside of the equipment, wire-bending space inside the equipment, and clearances between bare conductive parts on the inside of the equipment. The *NEC®* sets minimum requirements associated with these issues.

NEC Article 110, Part II includes minimum requirements for spaces about electrical equipment rated at 600 volts or less. Spaces and clearances covered in *NEC Article 110* include headroom (height), width, and depth of working spaces, as well as minimum requirements associated with entrances to working spaces associated with switchgear. Other space requirements covered in *NEC Article 110* include dedicated equipment space in indoor and outdoor installations, and spacing pertaining to the guarding of live parts.

NEC Article 408 covers regulations associated with the installation of switchboards and panelboards. Spacing requirements included in *NEC Article 408* include wire-bending space and minimum spacing between bare conductive metal parts.

In this project, the trainee will reference *NEC Articles 110* and *408* to determine the minimum spacing requirements associated with an equipment layout shown on a drawing.

OBJECTIVES

This performance project supports the following learning objectives listed in Module 26306-05:

- Describe switchgear construction, metering layouts, wiring requirements, and maintenance (*Objective 3*).
- List the *NEC®* requirements pertaining to switchgear (*Objective 4*).

PERFORMANCE TASKS

- There are no performance tasks associated with this module.

MATERIALS REQUIRED

- None

TOOLS AND EQUIPMENT REQUIRED

- Pencils and paper

REFERENCE MATERIALS AND LEARNING RESOURCES

- *NEC Articles 110* and *408*
- Module 26306-05, Sections 3.0.0, 3.4.0, 3.5.0, 4.0.0, 6.0.0, and 6.1.0

NOTES TO TRAINEE

- No safety equipment is required for this project unless the environment in which the project is completed requires safety equipment.
- Use the latest edition of the *National Electrical Code®* to complete this project.

NOTES TO INSTRUCTOR

- The instructor's solution is located at the end of this project.

PROCEDURE

This performance project requires you to review a switchgear drawing and determine the minimum spacing requirements based on given criteria and the *NEC®* minimum requirements listed in *NEC Articles 110* and *408*.

1. Carefully review the complete drawing in Figure 1.
2. Determine the minimum side working space (measurement A) based on the conditions shown in Figure 1. Write your answer in the space provided.
3. Determine the minimum headroom requirement (measurement B) based on the conditions shown in Figure 1. Write your answer in the space provided.
4. Determine the minimum spacing between busbars and busbars and ground based on the criteria given in Figure 1. Write your answer in the spaces provided.
5. Have your instructor check your work.

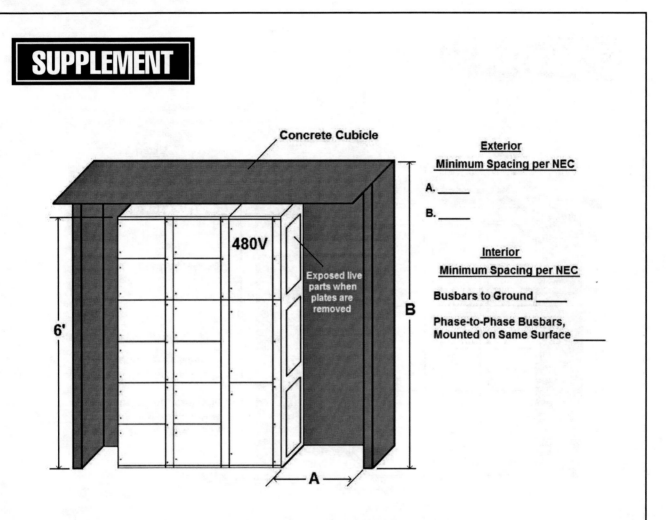

Concrete Cubicle

480V

Exposed live
parts when
plates are
removed

6'

B

A

Exterior
Minimum Spacing per NEC

A. _____

B. _____

Interior
Minimum Spacing per NEC

Busbars to Ground _____

Phase-to-Phase Busbars,
Mounted on Same Surface _____

Figure 1 ■ Switchgear Layout and Specifications

SOLUTION

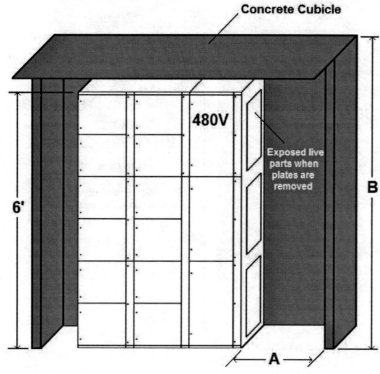

Concrete Cubicle

480V

Exposed live parts when plates are removed

6'

B

A

Exterior
Minimum Spacing per NEC

A. 3-1/2 feet
NEC Table 110.26(A)(1)
Condition 2

B. 6-1/2 feet
NEC Section 110.26(E)

Interior
Minimum Spacing per NEC

Busbars to Ground 1 inch
NEC Table 408.56

Phase-to-Phase Busbars,
Mounted on Same Surface 2 inches
NEC Table 408.56

Name_____ Date _____

PROJECT OVERVIEW

Distribution equipment testing and maintenance involves many skills and procedures. Electricians and technicians who are responsible for electrical maintenance on distribution equipment must know how to use various types of test equipment. One specialized piece of electrical test equipment that is directly associated with maintaining electrical distribution equipment is the megger.

A megger is used to test the insulating integrity or resistance of electrical conductors including wires, cables, and busbars. A simple continuity tester or ohmmeter applies a very low voltage and current through the conductors to test for continuity. A megger, on the other hand, generates a relatively high voltage at a very low current flow through the conductors. Any weakened spot in insulation or narrowed clearance between conductors or ground will cause the current to be pushed by the high voltage to the grounded part or other phase conductor, depending on where the test leads are connected. Meggers are available in a range of voltage capacities and are selected based on the system voltage that will be applied to the conductors.

In this project, the trainee will use a 1,000-volt megger to perform an insulation test on a de-energized panelboard.

OBJECTIVES

This performance project supports the following learning objectives listed in Module 26306-05:

■ Describe the visual and mechanical inspections and electrical tests associated with low-voltage and medium-voltage cables, metal-enclosed busways, and metering and instrumentation (*Objective 5*).

PERFORMANCE TASKS

■ There are no performance tasks associated with this module.

MATERIALS REQUIRED

■ One 600-volt maximum, single-phase or three-phase panel (installed or off-the-shelf)
■ Panelboard manufacturer's guidelines on insulation resistance

TOOLS AND EQUIPMENT REQUIRED

■ Pencils and paper
■ 1,000-volt megger

Module 26306-05

**Project 6-2: Perform an Insulation
Resistance Test Using a Megger**

REFERENCE MATERIALS AND LEARNING RESOURCES

- Module 26306-05, Sections 5.0.0, 5.1.0, and 5.2.0

NOTES TO TRAINEE

- You must wear safety glasses with side shields and 500-volt insulating gloves during this project.
- Do not attempt any testing on any panel until your instructor tells you to proceed. Insulation testing is NEVER performed on an energized panelboard or switchboard, or with any circuit breakers in an ON position.
- Meggers produce dangerously high voltage and are not toys. Do not horseplay using meggers. Follow all safety procedures associated with the use of a megger and have your instructor instruct you on safely using the megger before performing this project.

NOTES TO INSTRUCTOR

- Acquire an uninstalled panelboard for this project if possible.
- If this project is to performed on an existing installed panel, make sure the panel's feeder circuit and any backfeed circuits are locked out and tagged out. It is the instructor's responsibility to establish a safe working condition on any existing panel before allowing trainees to begin this project.
- All circuit breakers installed in an existing panel must be turned to the OFF position before performing the insulation testing. Follow the steps in Module 26306-05.
- Discourage trainee horseplay when using meggers.
- Instruct the trainees on the proper and safe use of meggers before beginning this project.
- The big issue with this project is safety. It is up to the instructor to teach megger safety before the project is started, and to maintain safe work practices throughout the project.

PROCEDURE

This performance project requires you to use a megger to perform an insulation test on a panelboard.

1. Have your instructor demonstrate to you the safe and correct use of the megger before beginning this project.
2. Verify with your instructor that the panelboard has been put in a safe work condition according to OSHA standards.

continued

PERFORMANCE PROJECTS

3. Perform an insulation resistance test on each bus section (phase-to-phase and phase-to-ground) for one minute.
4. Refer to the manufacturer's specific guidelines pertaining to insulation resistance, or refer to the table in Figure 1.
5. Write your results in the spaces provided on Figure 1.
6. Have your instructor check your work.

Phase-to-Phase Phase-to-Ground	Minimum Test (Megger) Voltage	Recommended Minimum Insulation Resistance (Megohms)	Actual Reading (Megohms)
Phase A to B	1,000	100	
Phase B to C	1,000	100	
Phase C to A	1,000	100	
Phase A to Ground	1,000	100	
Phase B to Ground	1,000	100	
Phase C to Ground	1,000	100	

Maximum 600-Volt Panel

Figure 1 ■ Insulation Resistance Test Table

ELECTRICAL
PERFORMANCE PROJECT

NOTES

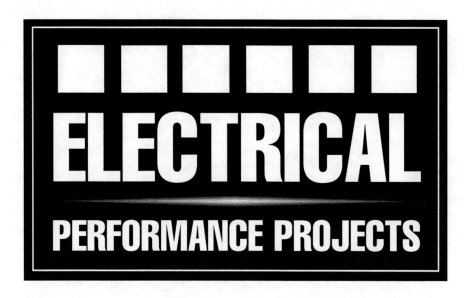

MODULE 26307-05

Projects:

- ■ **7-1 Estimate the Secondary Voltage When Connecting the Primary Winding of a 480V/120V Control Transformer to a 120-Volt Supply. Connect the Transformer and Measure the Secondary Voltage**

- ■ **7-2 Connect a Three-Phase Delta-Wye Transformer**

Project 7-1: Estimate the Secondary Voltage When
Connecting the Primary Winding of a 480V/120V
Control Transformer to a 120-Volt Supply. Connect the
Transformer and Measure the Secondary Voltage

Name_____ Date _____

PROJECT OVERVIEW

Transformers are used to increase or decrease a level of voltage. The source voltage is connected to the primary winding of the transformer and the load is connected to the secondary winding. The voltage available at the secondary winding is directly proportional to the ratio of the number of turns or coils between the two windings. If the primary winding has twice as many coils as the secondary winding, the transformer is said to have a ratio of two to one, and the secondary voltage available will be approximately half the value of the source voltage. On the other hand, if the primary winding has half as many coils or turns as the secondary winding, the ratio is one to two and the secondary voltage available will be approximately twice that of the source voltage.

Transformers that decrease voltage are called step-down transformers and transformers that increase voltage are called step-up transformers. Most transformers are connected as step-down transformers because when voltage is increased through a transformer, the amount of current available at the secondary winding is decreased in direct proportion to the ratio between the two windings. However, by decreasing the voltage through a transformer, the available current is increased in direct proportion to the ratio between the two windings.

In this project, the trainee will estimate and then measure the secondary winding voltage of a control transformer when the primary winding of a 480V/120V control transformer is connected to a 120-volt supply instead of a 480-volt supply. The purpose of the project is to demonstrate how the ratio of the windings is still maintained regardless of the voltage supply level.

OBJECTIVES

This performance project supports the following learning objectives listed in Module 26307-05:

- Describe transformer operation (*Objective 1*).
- Describe the operating characteristics of various types of transformers (*Objective 3*).
- Connect a control transformer for a given application (*Objective 8*).

PERFORMANCE TASKS

This performance project supports the following performance task(s) listed for Module 26307-05:

- Sketch or physically connect a dual-voltage transformer showing a high primary and a low secondary, including proper bonding (*Task 1*).

Module 26307-05

Project 7-1: Estimate the Secondary Voltage When Connecting the Primary Winding of a 480V/120V Control Transformer to a 120-Volt Supply. Connect the Transformer and Measure the Secondary Voltage

MATERIALS REQUIRED

- One control transformer, 480-volt primary, 120-volt secondary
- One 120-volt pigtail with cap (plug)

TOOLS AND EQUIPMENT REQUIRED

- Screwdriver set
- Wire cutters
- Wire strippers
- Digital multimeter

REFERENCE MATERIALS AND LEARNING RESOURCES

- Module 26307-05, Section 2.0.0, 2.1.0, 2.2.0, and 2.3.0

NOTES TO TRAINEE

- Safety glasses with side shields (or goggles) must be worn during this project.
- Even though the transformer is designed for 480 volts on the primary winding, we will connect its primary winding to a 120-volt supply. There will be no 480-volt connections made to this transformer.

WARNING!
Control step-down transformers that are reverse-connected (source to secondary winding) will produce dangerously high voltage levels on the opposing winding because the transformer acts as a step-up transformer. Use a digital ohmmeter to identify the primary and secondary windings if the terminals are not clearly marked. On a step-down transformer, the primary winding will have more resistance than the secondary winding because it has more turns of wire. Verify this before making connections on unmarked or unclearly marked transformers.

- Use a 120-volt pigtail to energize the primary side of the transformer.
- Wear 500-volt rated gloves when measuring voltage at the secondary winding.

NOTES TO INSTRUCTOR

- The ratio of the control transformer should be approximately 4:1 (480V/120V), and connected to a supply voltage of 120 volts.
- The transformer will be connected as a step-down or voltage reducing transformer.
- The primary side of the transformer will be connected to 120-volt supply through a 120-volt pigtail.

continued

Module 26307-05

Project 7-1: Estimate the Secondary Voltage When Connecting the Primary Winding of a 480V/120V Control Transformer to a 120-Volt Supply. Connect the Transformer and Measure the Secondary Voltage

ELECTRICAL

PERFORMANCE PROJECT

- One purpose of this project is to demonstrate to trainees that even though the transformer is labeled as 480V/120V, the voltage ratio is maintained when connected to a lower-than-listed primary voltage.
- Enforce safe work practices by having trainees wear safety glasses with side shields (or goggles) and 500-volt rated gloves when taking voltage measurements.

PROCEDURE

This performance project requires you to connect and measure the secondary voltage of a 480V/120V (4:1 ratio) control transformer that is connected to a 120-volt supply instead of a 480-volt supply.

1. Verify that no power is connected to the control transformer.
2. Set your digital multimeter to measure ohms and measure the resistance of both windings.
3. The winding with the higher resistance is the primary side and will be connected (not yet!) to the 120-volt supply.
4. The winding with the lower resistance is the secondary side and will not be connected to anything.
5. Notice that the resistance is not directly proportional to the ratio of the windings. The ratio of the windings for this transformer should be 4:1, but the ratio of the two resistance values is greater than 4:1. Many factors such as load, heat, current flow and others affect the resistance ratio.
6. Connect the terminal end of an unplugged 120-volt pigtail to the primary side of the transformer.
7. Make sure your safety glasses are in place. Plug the 120-volt pigtail into a standard 120-volt receptacle.
8. Estimate the voltage on the secondary side of the transformer, based on the ratio, and write your answer in the space provided on Figure 1.
9. Set your digital multimeter to measure AC voltage, put on your voltage-rated gloves, and measure the secondary voltage.
10. Write the actual voltage in the space provided on Figure 1.
11. Have your instructor check your work.
12. De-energize your transformer.

Module 26307-05

Project 7-1: Estimate the Secondary Voltage When Connecting the Primary Winding of a 480V/120V Control Transformer to a 120-Volt Supply. Connect the Transformer and Measure the Secondary Voltage

SUPPLEMENT

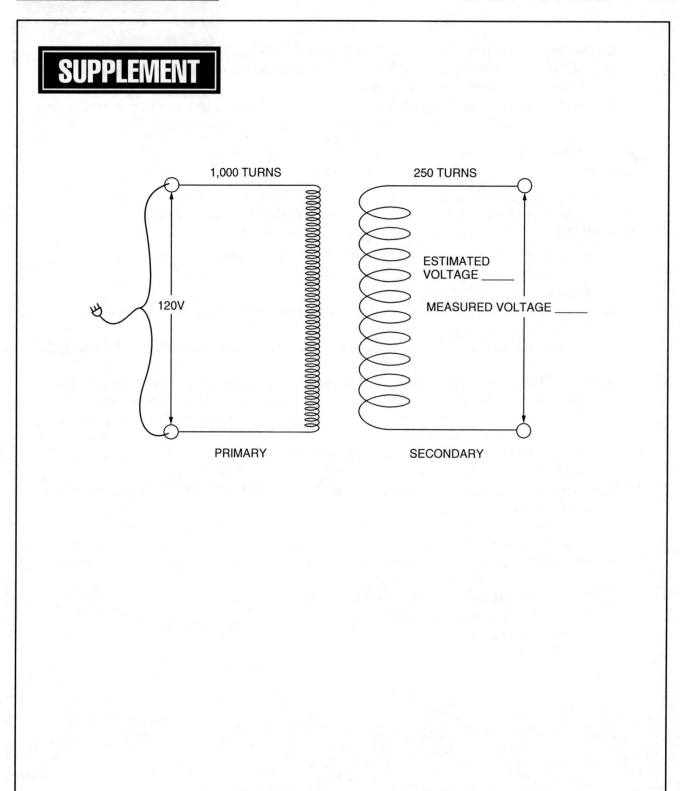

1,000 TURNS

250 TURNS

120V

ESTIMATED VOLTAGE _____

MEASURED VOLTAGE _____

PRIMARY

SECONDARY

Name_____ Date _____

PROJECT OVERVIEW

A common transformer system found in most commercial and industrial locations is the delta primary-wye secondary transformer system. As a three-phase, four-wire system, it provides three phases of power for equipment and single-phase power for lighting and other single-phase loads.

A delta-wye transformer system is three separate single-phase, step-down transformers interconnected to one another on both the primary and secondary sides. However, three-phase transformers are not created by simply field-connecting any three single-phase transformers together. These systems are internally wired and manufactured in a factory environment, designed and rated to operate on a specific primary voltage, and supply a specific load connected to the secondary winding.

Electricians should understand the internal wiring of a delta-wye transformer and be able to connect the primary and secondary conductors to them.

This project gives the trainee the opportunity to diagram the internal and field connections of a delta-wye transformer.

OBJECTIVES

This performance project supports the following learning objectives listed in Module 26307-05:

- Describe transformer operation (*Objective 1*).
- Explain the principle of mutual induction (*Objective 2*).
- Describe the operating characteristics of various types of transformers (*Objective 3*).
- Connect a multi-tap transformer for the required secondary voltage (*Objective 4*).
- Explain types and purposes of grounding transformers (*Objective 7*).

PERFORMANCE TASKS

This performance project supports the following performance task(s) listed for Module 26307-05:

- Sketch or physically connect a three-phase transformer in a delta-wye and open delta configuration, including proper bonding (*Task 3*).

MATERIALS REQUIRED

- Pencils
- Multiple copies of Figure 1

TOOLS AND EQUIPMENT REQUIRED

- None

REFERENCE MATERIALS AND LEARNING RESOURCES

- Module 26307-05, Sections 2.0.0, 5.0.0, and 5.2.0

NOTES TO TRAINEE

- No safety equipment is required for this project unless the environment in which the project is completed requires safety equipment.
- Make sure you draw in the grounding connection in the proper place on the secondary side of the transformer.
- You should first reference Figure 15 in Module 26307-05, then perform this project with your book closed.

NOTES TO INSTRUCTOR

- Trainees should be able to complete this project without using their modules.
- Verify that the trainee draws in the grounding connection on the secondary (wye) side of the transformer.
- The instructor's solution is located at the end of this project.

PROCEDURE

This performance project requires you to illustrate your understanding of three-phase, delta-wye transformer systems by completing both the internal and external connections of the transformer system on a drawing.

1. Briefly review Module 26307-05, Figure 15 in your module, then close your book.
2. Complete the schematic drawings for the connections on Figure 1 by interconnecting the dots representing terminal points.
3. Label the nominal voltages available on the secondary side of the transformer.
4. Complete the hardwire drawings on Figure 1 by interconnecting the illustrated transformer connections.
5. Have your instructor check your work.

SUPPLEMENT

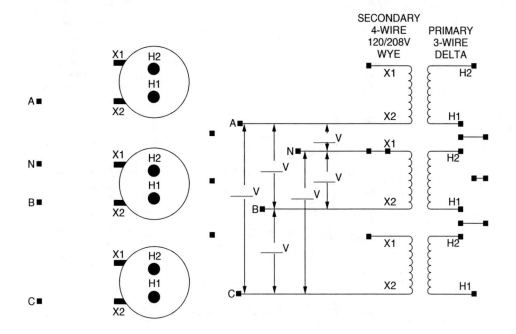

HARDWIRE ILLUSTRATION   **SCHEMATIC ILLUSTRATION**

NOTE: DON'T FORGET THE GROUNDING CONNECTIONS.

Figure 1 ■ Delta-Wye Connection Illustration

SOLUTION

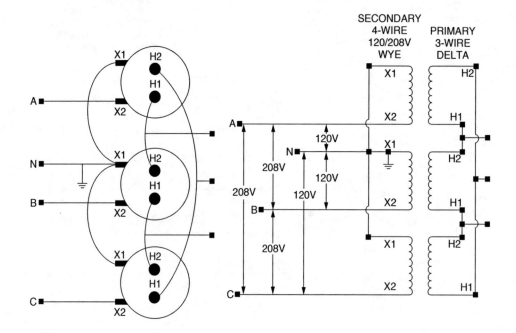

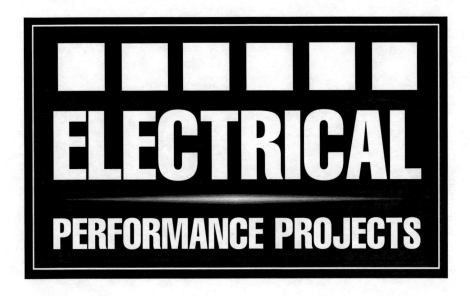

MODULE 26308-05

Projects:

- ■ **8-1** Selecting and Replacing an Instant-Start Ballast and Lamps
- ■ **8-2** Assembling and Testing a Photoelectric Lighting Fixture

ELECTRICAL

PERFORMANCE PROJECT

Module 26308-05
Project 8-1: Selecting and Replacing an
Instant-Start Ballast and Lamps

Name_____ Date _____

PROJECT OVERVIEW

Installing and maintaining electrical lighting is high on the list of tasks performed by electricians. Fluorescent lighting is a common type of lighting found in most commercial and some residential locations. Fluorescent lamps are low-pressure mercury discharge lamps that are energy efficient. They have a relatively long service life as compared to incandescent lighting.

A major component in fluorescent lighting is the ballast. The ballast provides a voltage that establishes an arc between two electrodes in the fluorescent lamp to start the lamp. In addition, the ballast regulates the electric current that flows through the lamp and stabilizes the light output.

Fluorescent ballasts are designed to operate in three different applications: preheat, rapid-start, and instant-start. Preheated fluorescent light fixtures use a separate starter switch that regulates the flow of preheating current through the lamps. These types of fixtures were popular during the early developmental stages of fluorescent lighting, but have now been mostly replaced by rapid- or instant-start fixtures. In rapid-start fixtures, there is no separate starter switch, and the ballast handles all preheating current flow control. Fluorescent lamps used in rapid-start fixtures are designed with two pins on each end of the lamp. Instant-start fluorescent fixtures are an improved variation of rapid-start fixtures. There is no preheating required in this type of lamp because the ballast provides a very high voltage across the electrodes, which causes electrons to be emitted from the electrodes. Instant-start fluorescent lamps are designed with a single pin on each lamp end because no preheating circuit is required.

In this project, the trainee will replace an instant-start ballast and two lamps in an instant-start fluorescent fixture.

OBJECTIVES

This performance project supports the following learning objectives listed in Module 26308-05:

- Recognize incandescent, fluorescent, and high-intensity discharge (HID) lamps and describe how each type of lamp operates (*Objective 1*).
- Recognize ballasts and describe their purpose for use in fluorescent and HID lighting fixtures (*Objective 2*).
- Use troubleshooting checklists to troubleshoot fluorescent and HID lamps and lighting fixtures (*Objective 5*).

ELECTRICAL LEVEL THREE

83

PERFORMANCE TASKS

This performance project supports the following performance task(s) listed for Module 26308-05:

- Use troubleshooting checklists and guidelines to troubleshoot selected fluorescent and/or HID lamps and lighting fixtures (*Task 1*).

MATERIALS REQUIRED

- Project board (plywood sheet or equivalent)
- 4-foot T8 instant-start fixture with ballast
- Replacement instant-start ballast compatible with fixture
- Replacement T8 fluorescent lamps compatible with fixture
- 120-volt pigtail with cap (plug) to power fixture
- ½" cable clamp to secure 120-volt pigtail at fixture
- Crimp-type ring terminal sized for the pigtail grounding conductor
- Six yellow wirenuts
- Sheet metal or wood screws for mounting fixture to board

TOOLS AND EQUIPMENT REQUIRED

- Pencil and paper
- Screwdriver set
- Wire strippers
- Wire cutters
- Lineman pliers
- Crimping pliers
- Slip-joint pliers

REFERENCE MATERIALS AND LEARNING RESOURCES

- Module 26308-05, Sections 4.0.0, 7.0.0, 7.1.0, 7.1.3, and 7.1.4

NOTES TO TRAINEE

- Goggles or safety glasses with side shields must be worn throughout this project.
- Work gloves must be worn when handling sharp metal fluorescent fixtures.
- Only work on de-energized fixtures.
- Handle fluorescent lamps with extreme care.

continued

- Check for pinched wires between the fixture housing and wiring cover before energizing the fixture.
- Verify that the replacement ballast and lamps are compatible with the fixture.

NOTES TO INSTRUCTOR

- The type of fluorescent fixture may be substituted based on availability.
- Check all wiring before the trainee installs the wiring cover.
- Encourage the wearing of safety glasses and gloves throughout the project.

PROCEDURE

This performance project requires you to replace an instant-start ballast and lamps in a fluorescent fixture.

1. Refer to the wiring schematic in Figure 1.
2. Remove the lamps (if installed) and wiring cover from the fluorescent fixture.
3. Use your lineman pliers to remove a ½" knockout plug from the fluorescent fixture.
4. Install the cable clamp and 120-volt pigtail onto the fixture. Do not energize the fixture at this time.
5. Secure the fixture to the project board.
6. Draw a simple ballast wiring connection diagram of the existing wiring, paying close attention to wiring colors.
7. Cut the wires from the ballast to the lamp sockets.
8. Remove the existing ballast and install the replacement ballast.
9. Connect the lamp socket wires to the ballast.
10. Prepare and connect the black and white 120-volt supply conductors to the line conductors of the ballast.
11. Install the crimp-type ring terminal onto the 120-volt pigtail grounding conductor and connect the terminal to the grounding screw on the fixture housing.
12. Check all wiring connections and have your instructor approve your wiring.
13. Install the wiring cover.
14. Install the lamps.
15. Energize the fixture and check your work.

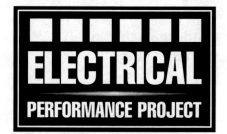

SUPPLEMENT

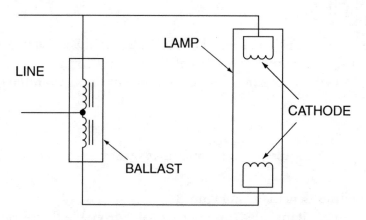

Figure 1 ■ Instant-Start Wiring Schematic

ELECTRICAL

PERFORMANCE PROJECT

Name_____ Date _____

PROJECT OVERVIEW

Outside lighting fixtures controlled by a photoelectric relay can be purchased as a complete unit, or they may be built from individually purchased components. One noteworthy fact about photocells is that they are not simple switches, but are actually relays with a coil and a set or sets of contacts. Simple photocells used in residential lighting control generally are designed with three conductors (black, white, and red) extending from the device. This often confuses inexperienced electricians who are expecting only two switch wires.

Photocells require a 120-volt supply to be connected to the black and white wires that extend from the photocell to supply power to the relay's coil. The red wire from the photocell is the switch leg that connects to black wire on the lighting fixture, along with a white neutral wire to each fixture.

This project allows the trainee to build and test a photoelectric lighting fixture by mounting lighting sockets referred to as PAR fixtures onto a cast aluminum round weatherproof box, installing a photocell, 120-volt pigtail supply cord, and interconnecting the PAR fixture wiring, photocell, and 120-volt supply conductors.

OBJECTIVES

This performance project supports the following learning objectives listed in Module 26308-05:

- Recognize incandescent, fluorescent, and high-intensity discharge (HID) lamps and describe how each type of lamp operates (*Objective 1*).
- Recognize ballasts and describe their purpose for use in fluorescent and HID lighting fixtures (*Objective 2*).
- Recognize basic occupancy sensors, photoelectric sensors, and timers used to control lighting circuits and describe how each device operates (*Objective 4*).
- Use troubleshooting checklists to troubleshoot fluorescent and HID lamps and lighting fixtures (*Objective 5*).

PERFORMANCE TASKS

This performance project supports the following performance task(s) listed for Module 26308-05:

- Use troubleshooting checklists and guidelines to troubleshoot selected fluorescent and/or HID lamps and lighting fixtures (*Task 1*).

MATERIALS REQUIRED

- See Figure 1.
- Project board (plywood sheet or equivalent)
- Round aluminum bell box with at least one ½" threaded conduit hub.
- Cover plate with three ½" threaded hubs.
- ½" cable clamp
- 120-volt pigtail with cap (plug)
- 120-volt photocell with ½" stem-mount connection and ½" locknut
- Two pre-wired PAR light sockets with ½" stem-mount connections and ½" locknuts
- Two incandescent 120-volt PAR lamps
- Six yellow wirenuts
- Sheet metal or wood screws for mounting box to board

TOOLS AND EQUIPMENT REQUIRED

- Screwdriver set
- Wire cutters
- Wire strippers
- Lineman pliers
- Slip-joint pliers

REFERENCE MATERIALS AND LEARNING RESOURCES

- Module 26308-05, Sections 2.0.0, 8.0.0, and 9.2.0

NOTES TO TRAINEE

- Safety glasses must be worn when working with conductors and lamps and when plugging in the 120-volt pigtail.
- Work gloves should be worn when working with sharp metal fixtures.
- Use extreme care when handling glass incandescent lamps.
- Handle the photocell carefully to avoid damage.
- Refer to the wiring diagram before connecting wires. Incorrect wiring can damage or destroy the photocell.
- Do not energize the fixture until approved by your instructor.
- Have your instructor check your wiring before installing the cover on the box.

NOTES TO INSTRUCTOR

- ■ The type of lamp fixtures or box may be substituted based on availability.
- ■ Check all wiring before allowing the trainee to install the box cover.

PROCEDURE

This performance project requires you to build a photoelectric fixture using one aluminum bell box, two PAR fixture sockets, and one photoelectric cell.

1. Obtain all the materials and tools listed in the material and tool lists (Figure 1).
2. Thread the ½" mounting stems of the two pre-wired PAR fixture sockets into the two outer ½" cover plate hubs and secure them in place with the locknuts provided with the fixtures.
3. Thread the ½" cable clamp into one ½" box hub.
4. Install the 120-volt pigtail through the cable clamp, into the box, and secure the clamp around the cable.
5. Thread the ½" mounting stem of the photocell into the center ½" hub on the box cover plate and secure it in place with the locknut provided.
6. Prepare all wire ends by stripping them to the correct lengths.
7. Use wirenuts to connect the wiring as shown in the wiring diagram of Figure 2.
8. Have your instructor check your work.
9. Install the cover on the bell box.
10. Energize the fixture with the approval of your instructor.
11. The lamps should immediately light. This is characteristic of photocell operation. If light is directed onto the photocell lens, the lamps should extinguish within five minutes of energizing.
12. Block the photocell lens from light to test its operation. The lamps should again light.
13. Continue testing the operation of the photocell until you and your instructor are satisfied with its operation.

SUPPLEMENT

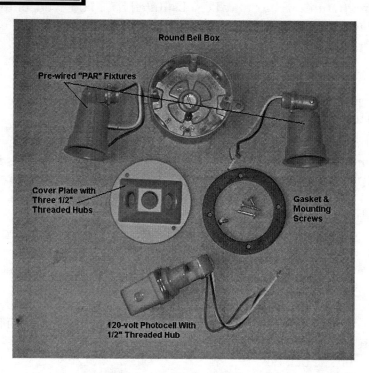

Figure 1 ■ Materials Required to Build a Photoelectric Lighting Fixture

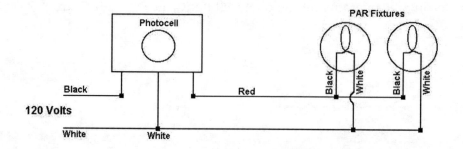

Figure 2 ■ Wiring Diagram

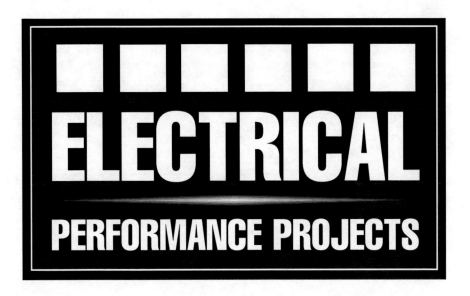

MODULE 26309-05

Projects:

Project 9-1: Sizing Circuit Protection and Conductors for a Branch Circuit Supplying Multiple Motors

Name_____ Date _____

PROJECT OVERVIEW

The *NEC®* permits two or more motors of any horsepower rating to be connected to a single branch circuit, as long as each individual motor is provided with overload protection and a short circuit protective device protects the branch circuit.

The *NEC®* limits the use of instantaneous trip circuit breakers in motor branch circuits to those types that are adjustable and are part of a combination motor controller providing coordinated motor overload protection. Circuit breakers not meeting the requirements associated with instantaneous trip devices, and installed to provide short circuit protection in a multiple motor application, are generally inverse time circuit breakers. These breakers incorporate a sufficient time-delay function in their operation to prevent nuisance tripping.

NEC Section 430.53(B) requires that the branch circuit overload protective device be sized no greater than the maximum value of short circuit protection, as shown in *NEC Table 430.52*, for the smallest motor in the group being served. In the case of inverse time-delay circuit breakers, this value equates to a maximum of 250 percent of the full load amperage of the smallest motor.

The time-delay characteristic of the circuit breaker, along with the 250 percent allowable rating, usually counters any problems associated with nuisance tripping due to increased start-up current flow.

NEC Section 430.24 states that when sizing branch circuit conductors supplying several motors, the branch circuit conductors must have an ampacity of not less than 125 percent of the full-load current rating of the highest rated motor plus the sum of the full-load current ratings of all the other motors.

In this project, the trainee will size a branch circuit short circuit protective device and branch circuit wiring for a circuit containing multiple motors.

OBJECTIVES

This performance project supports the following learning objectives listed in Module 26309-05:

- Size branch circuits and feeders for electric motors (*Objective 1*).
- Size motor short circuit protectors (*Objective 5*).
- Size multi-motor branch circuits (*Objective 6*).

PERFORMANCE TASKS

There are no performance tasks associated with this module.

MATERIALS REQUIRED

■ None

TOOLS AND EQUIPMENT REQUIRED

■ Pencils and paper
■ Calculator
■ Latest edition of the *National Electrical Code®*

REFERENCE MATERIALS AND LEARNING RESOURCES

■ *NEC Sections 240.4(D), 430.24, 430.40, 430.52, 430.53(B), Tables 310.16, 430.52,* and *430.250*
■ Module 26309-05, Sections 5.2.0 and 6.0.0

NOTES TO TRAINEE

■ No safety equipment is required for this project unless the environment in which the project is completed requires safety equipment.
■ Refer to *NEC Table 430.250* for all motor full-load amperage ratings.
■ *NEC 430.52(C)(1), Exception No. 1* permits using the next higher standard amperage rating for short circuit protective devices when the calculated amperage is not a standard rating for a circuit breaker. Refer to *NEC Section 240.6* for standard ratings.
■ Refer to *NEC Sections 430.24, 240.4(D)* and *Table 310.16* when sizing the branch circuit conductors.

NOTES TO INSTRUCTOR

■ Have the trainees read *NEC Sections 430.24* and *430.51* through *430.53(B)* before starting the project.
■ You may change the horsepower ratings of the motors for additional practice.
■ The solution is located at the end of this project.

PROCEDURE

This performance project requires you to determine the full-load amperage of given motors based on horsepower and *NEC Table 430.250*. In addition, you will calculate the maximum branch circuit short-circuit protection rating based on *NEC Table 430.52* and determine the minimum size branch circuit conductors based on *NEC Sections 240.4(D), 430.24,* and *Table 310.16*.

continued

Project 9-1: Sizing Circuit Protection and Conductors for a Branch Circuit Supplying Multiple Motors

1. Read *NEC Sections 430.24* and *430.51* through *430.53(B)*.
2. Refer to Figure 1 to complete this project.
3. Refer to *NEC Table 430.250* and find the full-load current based on horsepower for each motor shown in Figure 1.
4. Calculate the maximum value of short circuit protection (250 percent) of the smallest motor of the group shown in Figure 1.
5. If the calculated amperage in Step 4 is not a standard rating for circuit breakers, refer to *NEC Section 240.6* to determine the next higher standard rating. Write this value in the space provided on Figure 1.
6. Calculate branch circuit conductor ampacity. Refer to *NEC Section 430.24*.
7. Use *NEC Table 310.16* to locate the minimum size THHN copper conductors that may be installed in this application, based on the results of Step 6.
8. Have your instructor check your work.

SUPPLEMENT

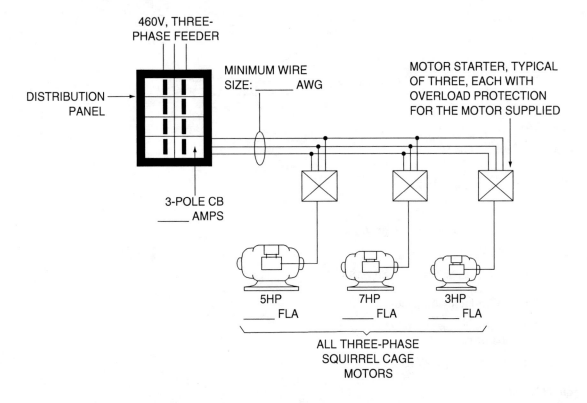

Figure 1 ■ Multiple-Motor Branch Circuit

SOLUTION

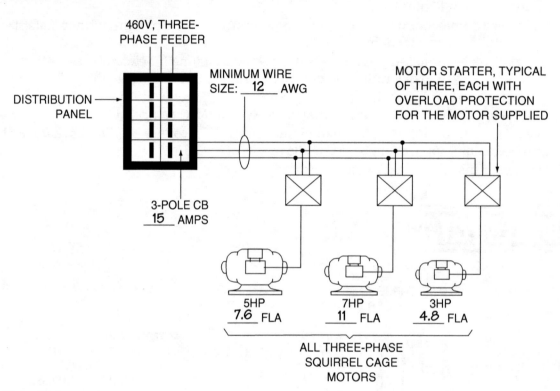

460V, THREE-PHASE FEEDER

MINIMUM WIRE SIZE: __12__ AWG

MOTOR STARTER, TYPICAL OF THREE, EACH WITH OVERLOAD PROTECTION FOR THE MOTOR SUPPLIED

DISTRIBUTION PANEL

3-POLE CB __15__ AMPS

5HP __7.6__ FLA

7HP __11__ FLA

3HP __4.8__ FLA

ALL THREE-PHASE SQUIRREL CAGE MOTORS

Motor FLAs:

7½ hp	11A
5 hp	7.6A
3 hp	4.8A

Short Circuit Protective Device:

250 percent of smallest motor amperage, or:

$2.50 \times 4.8 = 12A$

Next higher standard circuit breaker rating: 15A

Branch Circuit Conductor Ampacity:

125 percent of FLA of largest motor amperage + the sum of the rest of the FLAs, or:

$(1.25 \times 11) + (7.6 + 4.8) = 26.15A$

From *NEC Table 310.16*:

12 AWG THHN is good for 30 amperes

ELECTRICAL

PERFORMANCE PROJECT

Name_____ Date _____

PROJECT OVERVIEW

There are a variety of ways to protect motor circuits, including nontime-delay fuses, dual-element time-delay fuses, instantaneous-trip circuit breakers, and inverse time-delay circuit breakers.

When fuses are used as motor protective devices, the ampere rating of the fuse selected depends on whether the fuse is of the dual-element, time-delay type or the nontime-delay type. *NEC Table 430.52* lists the maximum ratings of these devices based on the type of device. In general, this table specifies that, for three-phase AC squirrel cage motors, dual element time-delay fuses used as short circuit/ground fault protection can be sized at 175 percent of the motor's full-load current rating, while nontime-delay fuses used for the same purpose can be sized at 300 percent of the motor's full-load current value.

Whenever these percentage values calculate to a non-standard fuse size as listed in *NEC Section 240.6*, the next higher standard fuse value can be used. Another exception to the general rule of fuse sizing for motor protection is that, if necessary to accommodate motor starting, the fuse size may be increased up to 400 percent of the full-load current of the motor.

The project provides practice in sizing both nontime-delay and dual-element time-delay fuses for several branch motor circuits supplied by a motor distribution panel.

OBJECTIVES

This performance project supports the following learning objectives listed in Module 26309-05:

■ Size branch circuits and feeders for electric motors (*Objective 1*).
■ Size motor short circuit protectors (*Objective 5*).

PERFORMANCE TASKS

There are no performance tasks associated with this module.

MATERIALS REQUIRED

■ None

TOOLS AND EQUIPMENT REQUIRED

■ Pencils and paper
■ Calculator
■ Current edition of the *National Electrical Code®*

REFERENCE MATERIALS AND LEARNING RESOURCES

■ *NEC Sections 240.6* and *430.52*, and *Tables 430.52* and *430.250*
■ Module 26309-05, Sections 4.0.0 and 4.1.0.

NOTES TO TRAINEE

■ No safety equipment is required for this project unless the environment in which the project is completed requires safety equipment.
■ Refer to *NEC Table 430.250* for all motor full-load amperage ratings.
■ Refer to *NEC Table 430.52* for maximum fuse ratings.
■ Refer to *NEC Section 240.6* for standard fuse ratings.

NOTES TO INSTRUCTOR

■ You may change the horsepower ratings of the motors for additional practice.
■ The instructor's solution is located at the end of this project.

PROCEDURE

This performance project requires you to size both dual-element time-delay and nontime-delay fuses for various branch motor circuits.

1. Look up and write down the full-load current in *NEC Table 430.250* for each of the motors shown on Figure 1.
2. Refer to *NEC Table 430.52* and determine the maximum fuse rating for each of the motors based on the type of fuse shown on Figure 1.
3. Calculate the fuse rating for each motor circuit based on the standard fuse ratings found in *NEC Section 240.6* and write your answers in the spaces provided on Figure 1.
4. Have your instructor check your work.

ELECTRICAL
PERFORMANCE PROJECT

SUPPLEMENT

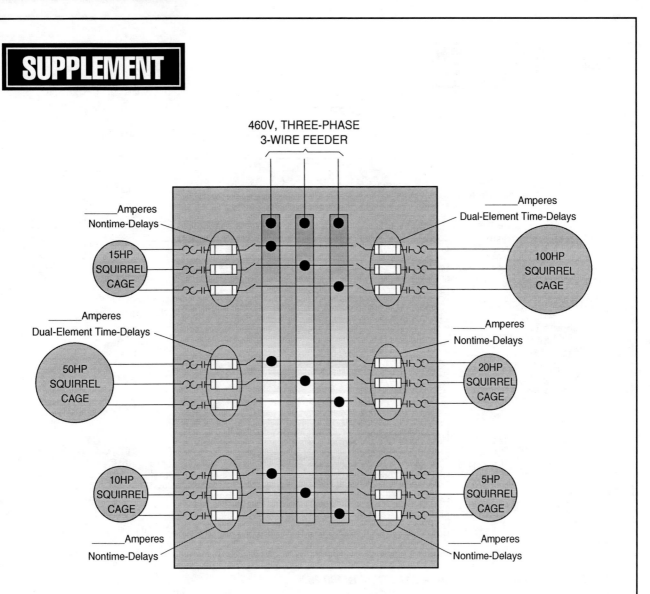

460V, THREE-PHASE 3-WIRE FEEDER

_____ Amperes Nontime-Delays

15HP SQUIRREL CAGE

_____ Amperes Dual-Element Time-Delays

100HP SQUIRREL CAGE

_____ Amperes Dual-Element Time-Delays

50HP SQUIRREL CAGE

_____ Amperes Nontime-Delays

20HP SQUIRREL CAGE

10HP SQUIRREL CAGE

_____ Amperes Nontime-Delays

5HP SQUIRREL CAGE

_____ Amperes Nontime-Delays

Figure 1 ■ Motor Distribution System Drawing

SOLUTION

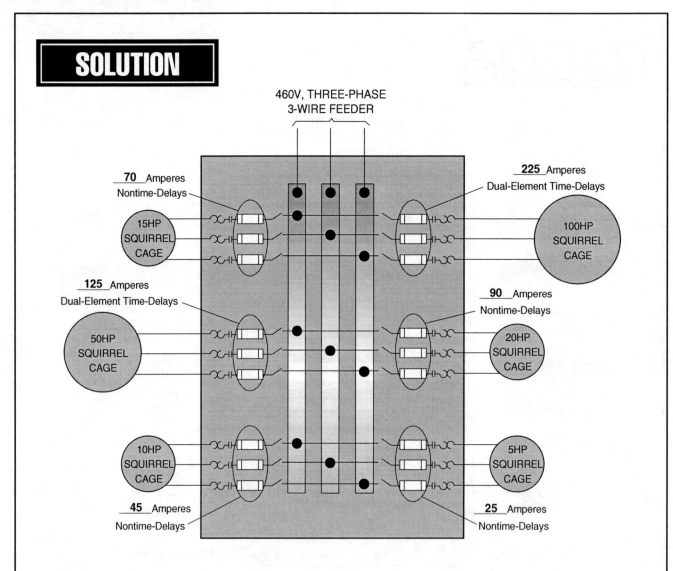

460V, THREE-PHASE
3-WIRE FEEDER

___70___ Amperes
Nontime-Delays

15HP
SQUIRREL
CAGE

___225___ Amperes
Dual-Element Time-Delays

100HP
SQUIRREL
CAGE

___125___ Amperes
Dual-Element Time-Delays

50HP
SQUIRREL
CAGE

___90___ Amperes
Nontime-Delays

20HP
SQUIRREL
CAGE

10HP
SQUIRREL
CAGE

5HP
SQUIRREL
CAGE

___45___ Amperes
Nontime-Delays

___25___ Amperes
Nontime-Delays

Individual Motor Full-Load Current (*NEC Table 430.250*):

15 hp	21A
50 hp	65A
10 hp	14A
100 hp	124A
20 hp	27A
5 hp	7.6A

Maximum Fuse Rating (*NEC Table 430.52*):

$21A \times 3.00$	$= 63$	Next Standard Fuse Size	70A
$65A \times 1.75$	$= 113.75$	Next Standard Fuse Size	125A
$14A \times 3.00$	$= 42$	Next Standard Fuse Size	45A
$124A \times 1.75$	$= 217$	Next Standard Fuse Size	225A
$27A \times 3.00$	$= 81$	Next Standard Fuse Size	90A
7.6×3.00	$= 22.8$	Next Standard Fuse Size	25A

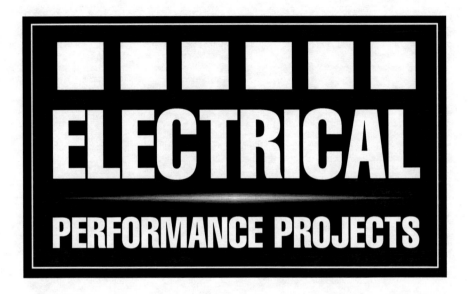

MODULE 26310-05

Projects:

- **10-1** Perform an Insulation Resistance Test on a
  Motor Using a Megger

- **10-2** Identify Unmarked Motor Leads on a
  Three-Phase, Nine-Lead Motor

Name_____ Date _____

PROJECT OVERVIEW

Many facilities do not wait for motors to experience problems before testing them. Preventive maintenance programs are commonplace in large industrial and commercial facilities that use a large number of motors, because motor replacements are expensive. Common motor maintenance programs may dictate replacing small single-phase motors and repairing larger three-phase motors. Regardless of the practice in place, part of motor maintenance and troubleshooting is knowing how to perform an insulation resistance test on a motor by using a megger.

Meggers are available in a range of high-voltage output levels. Before performing any insulation resistance test on a motor, a safety check must be completed to ensure that the circuits and equipment to be tested are rated to withstand the voltage level of the megger. Equipment and wiring should not be tested at voltage levels that greatly exceed their operating voltage. For example, equipment and wiring that normally operate at a voltage level in the range of 440 to 480 volts should be tested using a megger generating a 500-volt output, unless proprietary specifications call for a higher range of voltage testing.

Once the megger voltage level is determined, the megger should be tested for proper operation according to the megger manufacturer's recommendations or procedures. Before testing the motor, all motor T-leads should be disconnected from one another in the motor terminal box. Tag the wiring connections before disconnecting them. Each motor lead should be tested for continuity between the lead and the motor's frame or housing. The general rule is that the resistance reading between any one lead and the motor frame should not exceed one megohm, or one million ohms. Compensation or adjustment to this reading may be necessary based on the ambient temperature of the motor windings. The general rule is that for every 17° above 68°F, the megohm reading should be doubled to arrive at the actual reading. Likewise, for every 17° below 68°F, the reading should be halved. This follows the logic that as temperature increases, resistance decreases, and vice versa.

This project requires the trainee to use a megger to perform an insulation test on a motor.

OBJECTIVES

This performance project supports the following learning objectives listed in Module 26310-05:

- Test motors and generators (*Objective 1*).
- Select tools for motor maintenance (*Objective 7*).
- Select instruments for motor testing (*Objective 8*).

PERFORMANCE TASKS

This performance project supports the following performance task(s) listed for Module 26310-05:

- Using an untagged three-phase motor without a nameplate, determine as many motor characteristics as possible (*Task 1*).

MATERIALS REQUIRED

- One or more nine-lead, dual-voltage three-phase motors

TOOLS AND EQUIPMENT REQUIRED

- Pencils and paper
- Calculator
- Wire brush
- 500-volt megger
- Wire strippers
- Room thermometer

REFERENCE MATERIALS AND LEARNING RESOURCES

- Module 26310-05, Sections 4.0.0, 4.1.0, 4.2.0, 4.3.0, and 4.4.0

NOTES TO TRAINEE

- Goggles or safety glasses with side shields should be worn when using a wire brush, stripping wires, and a megger.
- Meggers produce high voltage levels. Do not touch the test leads while the megger is in use.
- Do not horseplay when using meggers. Meggers can cause serious shock.
- Never use a megger on an energized circuit or equipment.
- Wear work gloves when working with wire brushes, motors, and meggers.
- Do not adjust resistance readings if the ambient temperature of the motor winding is in a range of 52°F to 84°F.
- If the ambient temperature is not within this range, double the megohm reading for every 17°F above 68°F, or halve the megohm reading for every 17°F below 68°F.

NOTES TO INSTRUCTOR

■ Demonstrate proper usage of the megger before allowing trainees to use it.
■ Provide more than one motor, if possible, so that trainees may compare readings between the motors.
■ Personally test each motor prior to the start of the project and note the readings.

PROCEDURE

This performance project requires you to use a megger to perform a motor insulation resistance test, and calculate the adjusted resistance based on ambient temperature.

1. Have your instructor show you the proper and safe usage of the megger. Become familiar with its use.
2. Place the room thermometer near the motor and read the ambient temperature.
3. Record this temperature in the space provided on Figure 1.
4. Record a multiplication factor of 2 in the space provided on Figure 1 for every 17°F above 68°F.
5. Record a division factor of 2 in the space provided on Figure 1 for every 17°F below 68°F.
6. Prepare the motor T-leads by separating and exposing the bare ends of all nine leads in the motor termination box.
7. Locate and wire brush an area on the motor housing that will easily accommodate the alligator clip on the megger lead end.
8. Clip the megger lead alligator clip to the motor frame.
9. Locate T-lead #1 and clip the other megger test lead to T-lead #1.

WARNING!
Meggers produce high voltage levels. Do not touch the test leads or motor frame while the megger is in use.

10. Crank the megger handle (on manual models) or press the megger TEST button (on electronic models) for approximately one minute.
11. Record the reading in the space provided on Figure 1.
12. Repeat Steps 2 through 12 for T-leads 2 through 9.
13. Multiply or divide the readings by any factors entered in Steps 4 and 5, and enter the results in the space provided on Figure 1.
14. Remove the megger test leads.
15. Have your instructor check your work.

SUPPLEMENT

T-Leads	Reading	Ambient Temperature	Factor (÷ or ×)	Adjusted Reading
T1				
T2				
T3				
T4				
T5				
T6				
T7				
T8				
T9				

Figure 1 ■ Winding Resistance Test Record

ELECTRICAL

PERFORMANCE PROJECT

Name_____ Date _____

PROJECT OVERVIEW

One common task for industrial and commercial electricians is connecting nine-lead, three-phase motors for proper operation. Typically, all nine leads of a three-phase, dual-voltage motor are available and accessible in the motor's terminal box so that the motor may be connected for either the higher or lower voltage level. The electrician must be able to identify each of the leads when connecting a motor for operation. Most T-leads (motor leads) are pre-identified with either a crimp-on tag or a number imprinted directly on the motor lead insulation.

Unfortunately, these tags do not always remain in place, especially in situations where the motor has remained in service for a long time, or in some cases, after a motor has been rewound by a motor rewind shop. Untagged motors are usable motors and should not be discarded or left in a state of uncertainty. All electricians who regularly install and maintain three-phase motors must know how to identify unmarked motor leads, or at least be able to follow a step-by-step set of procedures to accomplish this task.

This project supplements the learning in Module 26310-05 and gives the trainee a hands-on opportunity to identify the internal circuits of a nine-lead three-phase motor.

OBJECTIVES

This performance project supports the following learning objectives listed in Module 26310-05:

- Test motors and generators (*Objective 2*).
- Make connections for specific types of motors and generators (*Objective 3*).
- Collect and record motor data (*Objective 6*).
- Select instruments for motor testing (*Objective 8*).

PERFORMANCE TASKS

This performance project supports the following performance task(s) listed for Module 26310-05:

- Using an untagged three-phase motor without a nameplate, determine as many motor characteristics as possible (*Task 1*).
- Sketch and/or physically connect the motor leads in the above motor for operation on its lower voltage (*Task 2*).
- Sketch and/or physically connect the motor leads of the same motor for operation on its higher voltage (*Task 3*).

MATERIALS REQUIRED

- One untagged (or tags concealed) nine-lead, dual-voltage three-phase motor (internally wound in a wye configuration)

TOOLS AND EQUIPMENT REQUIRED

- Pencils and paper
- Continuity tester or VOM

REFERENCE MATERIALS AND LEARNING RESOURCES

- Module 26310-05, Section 7.0.0

NOTES TO TRAINEE

- Safety glasses must be worn when working with motor leads.
- This project reflects step-by-step procedures outlined in Module 26310-05, Section 7.0.0 associated with identifying motor leads on an unmarked motor. However, the project only covers Steps 1 and 2, which identify the internal motor windings of a wye-connected motor. Steps 3 through 12 in your module continue with the process of identifying each of the motor leads by applying three-phase power to motor leads. These steps require the availability of three-phase power, and advanced electrical skills, and could expose you to dangerous shock hazards. For these reasons, only Steps 1 and 2 will be performed in this project. However, read through Steps 3 through 12 after completing Steps 1 and 2 to acquire a better understanding of the continued process.

NOTES TO INSTRUCTOR

- This project requires a motor with unmarked leads. If the T-leads of the available motor are marked, conceal the markings or numbers with tape or paint.
- The purpose of this project is to provide the trainee with a basic understanding of identifying motor T-leads. It partially follows the step-by-step procedures (Steps 1 and 2 only) presented in Module 26310-05, Section 7.0.0. Due to shock hazards associated with Steps 3 through 12 and the need for three-phase power, this project does not complete the motor lead identification process. It is essential to have your trainees read through Steps 3 through 12 after completing Steps 1 and 2, and provide a question and answer period based on these continuing steps.

PROCEDURE

This performance project requires you to use a continuity tester to identify the four internal circuits of a wye-wound, three-phase, dual-voltage, nine-lead motor.

1. Refer to Module 26310-05, Section 7.0.0.
2. Draw the motor coils to form a wye connection, as illustrated in Figure 1 of this project, and number the coils as shown.
3. At the motor leads, connect one probe of the continuity tester to any of the nine leads, and randomly check for continuity between that lead and each of the other eight leads.
4. If a reading is obtained between the connected lead and only one of the other eight leads, these two leads form one of the three two-wire circuits in the motor, such as 1 and 4, 2 and 5, or 3 and 8. Twist the ends of any identified two-wire circuit together for later identification.
5. If a reading is obtained between a lead and two other leads, these three wires form the three-wire circuit created by leads 7, 8, and 9. Twist the ends of the three-wire circuit together for later identification.
6. Continue this process until all three two-wire circuits and the one three-wire circuit are identified.
7. Read Steps 3 through 12 in Section 7.0.0 of your module to understand where you would normally go from this point to complete the motor lead identification process.
8. Have your instructor check your work.

SUPPLEMENT

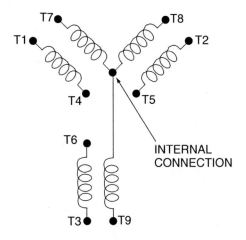

Figure 1 ■ Sketch Example of Wye-Connected Motor Windings

ELECTRICAL
PERFORMANCE PROJECT

NOTES

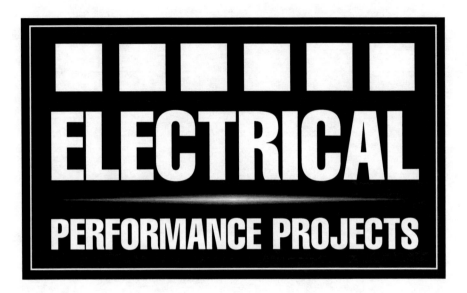

ELECTRICAL
PERFORMANCE PROJECTS

MODULE 26311-05

Projects:

- ■ **11-1 Install a Motor Control System Using Three-Wire Control**

- ■ **11-2 Disassemble and Reassemble a Magnetic Starter and Identify the Components**

ELECTRICAL

PERFORMANCE PROJECT

Name_____ Date _____

PROJECT OVERVIEW

Motors may be controlled in various ways, from a simple ON-OFF switch in small, single-phase motors to elaborate automated control systems involving programmable logic controllers and computers. Midway between the spectrum of motor control methods lies the three-wire remote motor control. Three-wire remote control uses two or more momentary-contact pushbuttons, commonly referred to as stop-start stations, to control energy to the coil on a magnetic motor starter. The terminology three-wire control is used because there are three control wires that must be installed between the field stop-start stations and the magnetic starter if the control power originates at the magnetic starter. If the control power originates from a location other than the magnetic starter (such as our project) only two wires are needed from the last stop-start station to the magnetic starter. When more than one stop-start station is installed in the same control circuit, three wires must also be installed between each station.

If control power is temporarily interrupted to a motor control ON-OFF switch incorporating maintained-contacts, the motor will immediately restart when power is restored to the control switch. On the other hand, if control power is temporarily interrupted to a three-wire momentary-contact control station, the START button must be pressed again to restart the motor. This is considered a positive personnel safety factor because motors are often temporarily stopped in an effort to clear something or someone from the operating functions of a motorized system.

Industrial electricians are frequently called upon to install and troubleshoot three-wire motor control systems. This project requires the trainee to install two three-wire control stations and interconnect them to operate a magnetic starter.

OBJECTIVES

This performance project supports the following learning objectives listed in Module 26311-05:

- Identify contactors and relays both physically and schematically, and describe their operating principles (*Objective 1*).
- Interpret motor control wiring, connection, and ladder diagrams (*Objective 3*).
- Connect motor controllers for specific applications according to *NEC®* requirements (*Objective 6*).

PERFORMANCE TASKS

This performance project supports the following performance task(s) listed for Module 26311-05:

- Make all connections for a magnetic motor controller, controlled by two pushbutton stations, including the connections for holding the circuit interlock (*Task 1*).

MATERIALS REQUIRED

- Project board (plywood sheet or equivalent)
- 120V/24V secondary-fused control transformer
- 120-volt pigtail with cap (plug) to energize transformer
- Magnetic starter (any NEMA size) with a 24V coil and a minimum of one set of latching contacts
- Two pushbutton stop-start stations
- 20 to 25 feet of red 16 AWG THHN copper wire
- 2 to 5 feet of white 16 AWG THHN copper wire
- Book of wire numbers (tags, labels, etc.)
- Screw-mount 6- or 8-inch cable ties (Tyraps® or equivalent)
- Sheet metal or wood screws for mounting stop-start stations, magnetic starter, and securing cable ties

TOOLS AND EQUIPMENT REQUIRED

- Screwdriver set
- Wire cutters
- Wire strippers
- Lineman pliers
- Marker
- Continuity tester

REFERENCE MATERIALS AND LEARNING RESOURCES

- Module 26311-05, Sections 3.0.0, 3.1.0, 3.3.0, 5.0.0, 5.2.0, 6.2.0, 6.2.1, and 6.2.5

NOTES TO TRAINEE

- Goggles or safety glasses with side shields must be worn when cutting and stripping wiring.
- Only 24 volts should be connected downstream from the transformer's secondary fuse block to the stop-start stations and magnetic starter. 120 volts should only supply the primary side of the transformer.
- Designate the fused transformer secondary conductor as L1 conductor.
- Designate the unfused transformer secondary conductor as L2 conductor.
- Run L2 directly from the transformer's secondary side to the L2 terminal on the magnetic starter.

continued

- The only wire terminated within the stop-start stations or latching contacts on the magnetic starter is L1.
- Avoid wild wire strands when terminating conductors to the switches and starter terminals.
- Use screw-mount cable ties and screws to neatly secure completed wiring bundles to the project board

NOTES TO INSTRUCTOR

- Make sure the secondary side of the transformer is fused with a maximum fuse rating of one ampere.
- Verify that L2 (unfused transformer secondary conductor) is wired directly from the transformer to the L2 termination point on the magnetic starter. This prevents a potential short circuit that may be caused by any mis-wiring of the stop-start stations.
- Stress neat workmanship in installing and securing the wiring.
- If the stop-start stations do not function as intended during the testing phase, de-energize the circuit and have the trainees troubleshoot their circuit using the drawing provided and a continuity tester.

PROCEDURE

This performance project requires you to install two momentary-contact stop-start stations and associated wiring that control the coil of a magnetic starter.

1. Mount the stop-start stations and the magnetic starter to the project board, similar to the layout shown in Figure 1.
2. Carefully review the schematic drawing in Figure 2.

WARNING!
Verify that the 120-volt supply pigtail to the transformer primary is disconnected from the AC supply.

3. Install a section of white wire from the unfused secondary side of the transformer to the L2 terminal on the magnetic starter. This terminal is usually located on the overload relay assembly.
4. Before wiring the stop-start stations, remove one knockout plug (if applicable) on each station enclosure to allow wire entry.
5. Install three pieces of red wire between Stations A and B, and two pieces from Station B to the magnetic starter.
6. Label each end of each wire between Station A and Station B as 1, 2, or 3, and the two wires from Station B as 2 and 3 on each end.

continued

7. Prepare the wire ends and terminate the wiring in Station A, Station B, and on the magnetic starter latching contact as shown in Figure 2.
8. Neatly secure wiring bundles to the project board using screw-mount cable ties and screws.
9. Have your instructor check your wiring before installing the station covers, then install the covers.
10. Energize the control circuit by plugging in the AC supply pigtail to a 120-volt receptacle.
11. Press the Start button on Station A. The starter coil should energize and the starter contactor pull in.
12. Press the Stop button on Station A and the starter coil should de-energize.
13. Repeat Steps 10 and 11 for Station B, looking for the same results.
14. If any switches do not respond as intended, unplug the circuit and troubleshoot your installation using a continuity tester and the schematic.

SUPPLEMENT

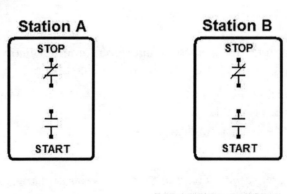

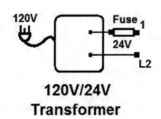

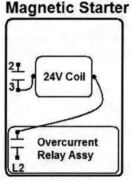

Figure 1 ■ Stop-Start Station and Magnetic Starter Physical Layout

SUPPLEMENT

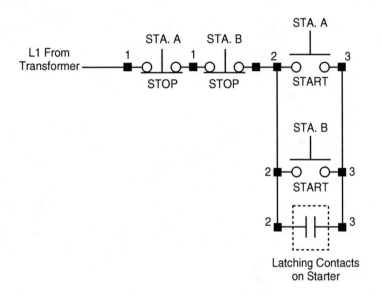

Figure 2 ■ Three-Wire Control Wiring Schematic

ELECTRICAL
PERFORMANCE PROJECT

NOTES

Module 26311-05
**Project 11-2: Disassemble and Reassemble a
Magnetic Starter and Identify the Components**

Name_____ Date _____

PROJECT OVERVIEW

A magnetic motor starter is a type of contactor that is used to control an electric motor. A magnetic motor starter contains many of the same components as a relay or lighting contactor, and some additional components.

A relay, contactor, and magnetic motor starter all contain a coil to operate the device, and one or more sets of main contacts that change state based on the state of the coil (energized or de-energized). The coil is electrically isolated from the main contacts and operates from an external control system. Contactors are nothing more than large relays handling larger current loads than most relays. The main contacts in any relay, contactor, or magnetic motor starter make or break circuit paths that are connected to them.

A magnetic motor starter usually contains an overload relay assembly that monitors the current flow level through the motor leads and opens a circuit if a current flow greater than a preset value occurs. Magnetic motor starters usually contain auxiliary sets of contacts that are used to control circuits in auxiliary devices such as pilot lights, interlocks, and other controlling or indicating devices. These auxiliary contacts are generally rated at a much smaller current value than the main contacts because they are only used to control small levels of current.

Electricians responsible for maintaining motor controls must know how to disassemble magnetic motor starters, identify components, replace components, and reassemble the starter to a working condition. This project affords the opportunity to disassemble and reassemble a typical magnetic motor starter.

OBJECTIVES

This performance project supports the following learning objectives listed in Module 26311-05:

- Identify contactors, relays both physically and schematically, and describe their operating principles (*Objective 1*).
- Interpret motor control wiring, connection, and ladder diagrams (*Objective 3*).
- Select and size contactors and relays for use in specific electrical motor control systems (*Objective 4*).
- Connect motor controllers for specific applications according to *NEC®* requirements (*Objective 6*).

Module 26311-05
Project 11-2: Disassemble and Reassemble a
Magnetic Starter and Identify the Components

PERFORMANCE TASKS

This performance project supports the following performance task(s) listed for Module 26311-05:

- Make all connections for a magnetic motor controller, controlled by two pushbutton stations, including the connections for holding the circuit interlock (*Task 1*).

MATERIALS REQUIRED

- One magnetic motor starter (Square D Class 8536, NEMA size 00 or equivalent)

TOOLS AND EQUIPMENT REQUIRED

- Pencils and paper
- Screwdriver set

REFERENCE MATERIALS AND LEARNING RESOURCES

- Module 26311-05, Sections 3.0.0, 3.1.0, 3.2.0, 3.2.2, 4.2.0, 5.0.0, 5.2.0, and 5.5.0

NOTES TO TRAINEE

- No safety equipment is required for this project unless the environment in which the project is completed requires safety equipment.
- Pay close attention and make notes when disassembling the magnetic starter so that you can reassemble it to its normal working condition.
- Identify and examine each part to see its function in the overall operation of the magnetic starter.
- You should be able to manually operate the movement of the movable main contact assembly after reassembly without any binding. If binding occurs, you must disassemble and reassemble the starter until you get it right.

NOTES TO INSTRUCTOR

- You may substitute the type of magnetic motor starter shown with any available starter, as long as the starter selected is electromechanical.
- Have the trainees manually operate the movement of the movable contact assembly before disassembly and after reassembly.

Module 26311-05
Project 11-2: Disassemble and Reassemble a
Magnetic Starter and Identify the Components

ELECTRICAL

PERFORMANCE PROJECT

PROCEDURE

This performance project requires you to disassemble a magnetic motor starter; identify primary components including the coil, main contacts (stationary and movable), and overload relay assembly; and reassemble the starter to an operable condition.

1. The following procedures reflect disassembly and reassembly of a Square D Class 8536, NEMA size 00 magnetic motor starter. If any other type of starter is used, your instructor will demonstrate the disassembly and reassembly procedures prior to starting the project.
2. Carefully examine the magnetic starter front and back in its whole condition, distinguishing between terminal screws and mounting screws (Figure 1).
3. Manually operate the movable contact assembly with a screwdriver blade, noting the direction of movement (Figure 2).
4. Disconnect any wires connected to the coil but do not fully remove them from the starter (Figure 3).
5. Remove the coil cover by loosening the two cover screws, exposing the coil (Figure 4).
6. Gently lift the coil from its housing, noting the removable iron core components inserted into the coil (Figure 5).
7. Separate the three coil parts, paying close attention to their orientation with one another (Figure 6).
8. Loosen the screws that hold the main contact cover in place, exposing the main contacts (Figure 7).
9. Gently remove the movable main contact assembly from its housing, paying close attention to its orientation with the housing. Note the stationary contacts still in place within the starter's housing (Figure 8).
10. Turn the starter assembly over to view the backside metal mounting plate. Locate the two small screw heads toward the bottom of the plate. Loosen but do not remove these screws (Figure 9).
11. Turn the starter assembly back to the front side and loosen the three recessed terminal screws directly above the overload relay assembly, but do not remove these screws (Figure 10).
12. Gently slide the relay assembly from the terminal screws (Figure 11).
13. Identify each component removed in the previous steps and review its function in the operation of the magnetic starter, then reassemble the starter.
14. Once reassembly is complete, repeat Step 3.

Module 26311-05
Project 11-2: Disassemble and Reassemble a
Magnetic Starter and Identify the Components

SUPPLEMENT

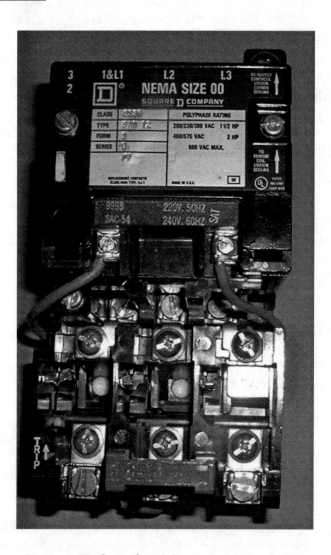

Figure 1 ■ Complete Magnetic Motor Starter

Module 26311-05
Project 11-2: Disassemble and Reassemble a
Magnetic Starter and Identify the Components

SUPPLEMENT

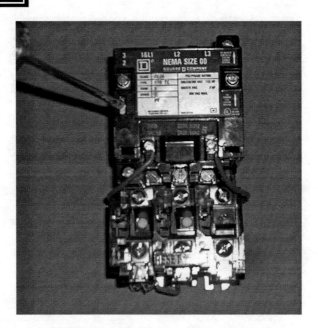

Figure 2 ■ Manual Operation of Movable Contacts

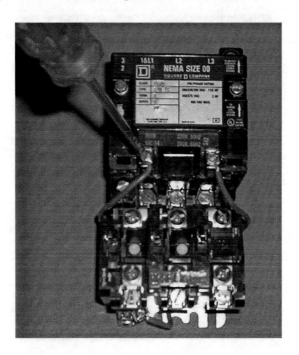

Figure 3 ■ Disconnecting Coil Wires

Module 26311-05
Project 11-2: Disassemble and Reassemble a
Magnetic Starter and Identify the Components

SUPPLEMENT

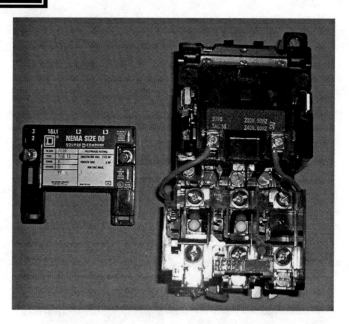

Figure 4 ■ Removing the Coil Cover

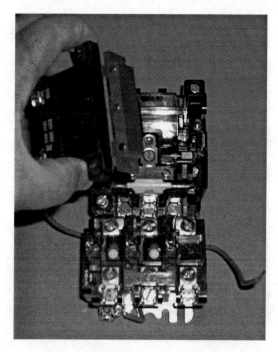

Figure 5 ■ Removing the Coil

Module 26311-05
Project 11-2: Disassemble and Reassemble a
Magnetic Starter and Identify the Components

SUPPLEMENT

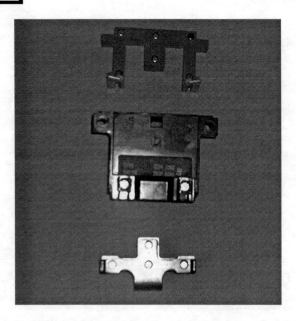

Figure 6 ■ Separating the Coil and Iron Core Parts

Figure 7 ■ Removing the Main Contact Cover

Module 26311-05
Project 11-2: Disassemble and Reassemble a
Magnetic Starter and Identify the Components

SUPPLEMENT

Figure 8 ■ Removing the Movable Contact Assembly

Figure 9 ■ Overload Relay Assembly Screws (Back of Starter)

ELECTRICAL
PERFORMANCE PROJECT

Module 26311-05
**Project 11-2: Disassemble and Reassemble a
Magnetic Starter and Identify the Components**

SUPPLEMENT

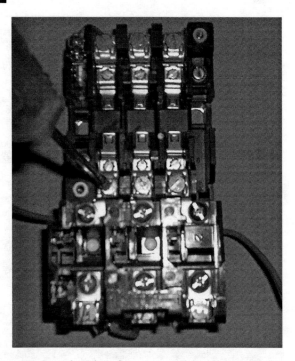

Figure 10 ■ Overload Relay Assembly Terminal Screws

ELECTRICAL
PERFORMANCE PROJECT

Module 26311-05
Project 11-2: Disassemble and Reassemble a
Magnetic Starter and Identify the Components

SUPPLEMENT

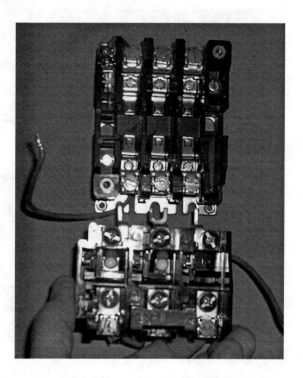

Figure 11 ■ Removing the Overload Relay Assembly

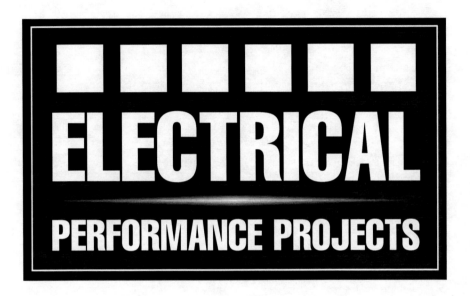

ELECTRICAL

PERFORMANCE PROJECTS

MODULE 26312-05

Projects:

- ■ 12-1 Installing Class One, Division One
  Conduit, Boxes, and Fittings

- ■ 12-2 Pulling Conductors Through and
  Filling a Sealing Fitting

PROJECT OVERVIEW

There are three hazardous classifications based on three general categories of ignitable or explosive materials. Each class has two divisions based on the condition or availability of the material. Class I involves ignitable or volatile gases or vapors, Class II covers ignitable or volatile dusts, and Class III includes fibers and flying materials that are considered ignitable. Division 1 for any class describes a condition in which the gas, vapor, dust, fibers, or flyings are present in the atmosphere in quantities that make the ambient condition hazardous. Division 2 identifies a condition in which the hazardous material is contained, and only becomes available in hazardous quantities if a fault occurs, such as a rupture in a tank, line, or any other abnormal failure or condition.

Electrical installations in hazardous locations must comply with *NEC®* regulations pertaining to both the class and the division. *NEC Chapter 5* contains all articles that cover electrical installations in hazardous locations.

Class I, Division 1 locations are those areas that contain volatile gases or vapors in the air in sufficient quantities to make the area hazardous. All electrical installations in these areas must comply with the *NEC®* regulations pertaining to Class I, Division 1 locations, which are covered in *NEC Article 501*. Conduit, fittings, boxes, devices, and fixtures installed in Class I, Division 1 locations must be approved for these locations and special wiring methods must be applied that comply with *NEC Article 501*.

In general, conduit installed in Class I, Division 1 locations must be either threaded rigid metal conduit or threaded steel intermediate metal conduit. All boxes and fittings must be approved for Class I, Division 1 locations.

In this project, the trainee will complete a partial Class I, Division 1 raceway installation using rigid conduit, boxes and fittings approved for Class I, Division 1 locations.

OBJECTIVES

This performance project supports the following learning objectives listed in Module 26312-05:

- Define the various classifications of hazardous locations (*Objective 1*).
- Describe the wiring methods permitted for branch circuits and feeders in specific hazardous locations (*Objective 2*).
- Select seals and drains for specific hazardous locations (*Objective 3*).
- Select wiring methods for Class I, Class II, and Class III hazardous locations (*Objective 4*).
- Follow *NEC®* requirements for installing explosionproof fittings in specific hazardous locations (*Objective 5*).

PERFORMANCE TASKS

This performance project supports the following performance task(s) listed for Module 26312-05:

■ Using two rigid metal conduit nipples, a sealing fitting, three pieces of 12 AWG THHN conductors, and a packing fiber/sealing kit, perform the following operations (*Task 1, Partial*):
   – Secure one conduit nipple in each end of the seal.
   – Make sure the required amount of threads is engaged.

MATERIALS REQUIRED

■ Project board (plywood sheet or equivalent)
■ ½" × 12" rigid conduit nipple
■ Two ½" rigid conduit close nipples
■ Sheet metal or wood screws (to anchor boxes and fittings)

Note:
   All of the following materials must be approved for Class I, Division 1 locations:

■ Two single-gang switch boxes with one ½" threaded hub on each
■ Two blank single-gang plates for switch boxes
■ UNF (female to female) ½" threaded union
■ ½" vertical sealing fitting

TOOLS AND EQUIPMENT REQUIRED

■ Two pair slip-joint pliers
■ Screwdriver set

REFERENCE MATERIALS AND LEARNING RESOURCES

■ *NEC Sections 501.10* and *501.15*
■ Module 26312-05, Sections 1.0.0, 1.1.0, 1.4.0, 2.0.0, 2.2.0, 3.0.0, 3.2.0, and 3.3.0

NOTES TO TRAINEE

■ Goggles or safety glasses with side shields and work gloves must be worn throughout this project.
■ A switch and wiring will be installed, and the sealing fitting filled in the next project. Do not fill it now.

continued

- Standard practice in the electrical trade is that the union nut should face down on vertical runs (nut portion union on top).
- The sealing fitting fill plug must be facing upward and toward the front of the installation.
- Conduit nipples, fittings, and boxes must be tightly joined together to protect the integrity of the hazardous location and keep the arcing device (switch) intrinsically secured from the ambient location.

NOTES TO INSTRUCTOR

- Types of conduit fittings and boxes may be substituted as long as they are approved for Class I, Division 1 locations.
- Verify that all connections are secure and not loosely joined. Emphasize an intrinsically safe installation.
- The seal will be filled in the next project.

PROCEDURE

This performance project requires you to install conduit nipples, fittings, a switch, and boxes in a simulated Class I, Division 1 location.

1. Acquire all materials listed in the material list.
2. Refer to Figure 1.
3. Use screws to mount one switch box (Box A) to the project board with the ½" conduit hub facing upward.
4. Thread one end of a ½" close nipple into the ½" hub on Box A.
5. Separate the conduit union into two parts and thread the end of the union that contains the nut threads (not the nut) onto the other end of the close nipple installed in Step 4. Tighten the close nipple and union half with slip-joint pliers.
6. Thread one end of another ½" close nipple into the bottom end of the sealing fitting.
7. Thread the other half of the union (with the union nut in place) onto the other end of the close nipple installed in Step 6. Tighten the close nipple union half and sealing fitting together with slip-joint pliers.
8. Thread one end of a 12" conduit nipple into the top end of the sealing fitting and tighten with slip-joint pliers.
9. Thread the assembly created in Steps 6 through 8 into the ½" hub of another switch box (Box B), making sure that the opening of the switch box and the sealing fitting fill plug align with one another, facing outward when tightened.
10. Secure the assembly created in Step 9 to the other half of the union fitting and join the two together by tightening the union nut.
11. Secure switch Box B to the project board with screws.
12. Install blank cover plates on both boxes.
13. Have your instructor check your work.

SUPPLEMENT

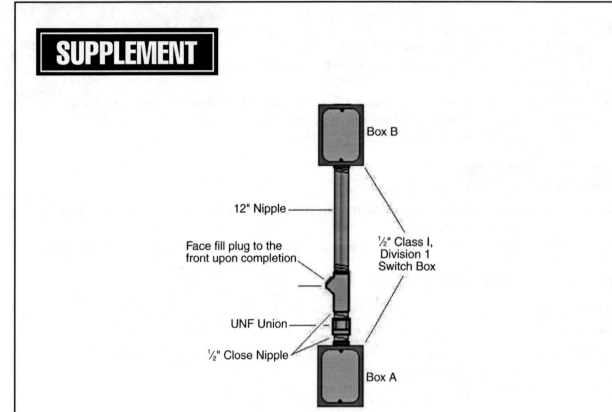

Figure 1 ■ Class I, Division 1 Partial Conduit Installation

ELECTRICAL

PERFORMANCE PROJECT

Name_____ Date _____

PROJECT OVERVIEW

The *NEC®* prohibits splices or taps in sealing fittings. After the conductors are pulled through a sealing fitting, the fitting must be filled with a sealing compound to prevent the passage of gases or vapors. When properly mixed and poured, the compound hardens into a dense, strong, solid mass that is insoluble in water, not affected by chemicals, and not softened or affected by heat. The compound is also designed to withstand the pressures associated with exploding trapped vapors or gases within the conduit system.

Seals should only be poured by electricians experienced in the methods of sealing. Seals should be poured according to the manufacturer's recommendations. One common brand of sealing compound is Chico®, although other brands of approved sealing compound are readily available.

In this project, the trainee will pull three conductors into the Class I, Division 1 conduit installation installed in the previous project, install a single-pole switch in Box A, then fill the sealing fitting with sealing compound.

OBJECTIVES

This performance project supports the following learning objectives listed in Module 26312-05:

- Describe the wiring methods permitted for branch circuits and feeders in specific hazardous locations (*Objective 2*).
- Select seals and drains for specific hazardous locations (*Objective 3*).
- Select wiring methods for Class I, Class II, and Class III hazardous locations (*Objective 4*).
- Follow *NEC®* requirements for installing explosionproof fittings in specific hazardous locations (*Objective 5*).

PERFORMANCE TASKS

This performance project supports the following performance task(s) listed for Module 26312-05:

- Using two rigid metal conduit nipples, a sealing fitting, three pieces of 12 AWG THHN conductors, and a packing fiber/sealing kit, perform the following operations (*Task 1, Partial*):
  - Pull the three THHN conductors through the nipples and seal so that about 6" is protruding from each nipple.
  - Pack the fiber as per the instructions furnished with the sealing kit.
  - Mix the sealing compound.
  - Position the unit in the required location and pour in the sealing compound.

MATERIALS REQUIRED

- Previously completed conduit installation, including conduit nipples, boxes, and sealing fitting.
- Nine feet of 12 AWG THHN wire (any color).
- Roll of black vinyl electrical tape.
- Chico® A4 sealing compound kit (or equivalent)
- Clean plastic quart-size mixing bucket or container (with pouring spout)
- Clean wooden or plastic spatula for mixing
- Cleaning rags

TOOLS AND EQUIPMENT REQUIRED

- Wire cutters
- Slip-joint pliers
- Screwdriver set
- Wrench for removing and installing sealing fitting fill plug

REFERENCE MATERIALS AND LEARNING RESOURCES

- *NEC Section 501.15(C)(2) and (3)*
- Module 26312-05, Section 3.3.4

NOTES TO TRAINEE

- Goggles or safety glasses with side shields and work gloves must be worn during the project.
- Make sure you install the conductors first.
- Follow the procedures given in Module 26312-05, Section 3.3.4 for installing the sealing kit fiber dam material and liquid compound.
- The purpose of the fiber dam is to prevent the liquid compound from flowing into the conduit and out of the sealing fitting.
- Do not overfill the fitting. You are only forming a seal. Refer to *NEC Section 501.15(C)(2) and (3)* for *NEC®* requirements.
- Allow the compound to set firmly before pulling or tugging on the conductors.
- Properly dispose of any remaining mixed compound.
- Clean any compound residue from the conduit or fittings.

NOTES TO INSTRUCTOR

- The conductors are installed in the conduit installation for simulation purposes only. They will not be used for anything beyond the experience of filling a sealing fitting with conductors installed, which is the normal situation.
- Check the installation of the fiber dam before allowing trainees to continue with the pouring process. It must be distributed around the conductors to prevent the liquid compound from flowing into the conduit.
- Do not allow the trainees to overfill the fitting.

PROCEDURE

This performance project requires you to install conductors in a partial conduit system containing a sealing fitting, and properly seal the fitting using an approved sealing kit.

1. Review Figure 1.
2. Acquire the partial conduit installation completed in the previous project.
3. Remove the blank covers from the Boxes A and B and the fill plug from the sealing fitting.
4. Cut three equal lengths (approximately 36" each) of 12 AWG THHN wire.
5. Tape one end of the three conductors together (no need to strip conductors) and push the taped end of wires from Box B to Box A, leaving equal lengths of wire at each box opening.
6. Roll and tuck the excess wire into the openings of each box and reinstall the blank box covers.
7. Open the sealing kit, read the instructions, and install the fiber dam in the sealing fitting.
8. Have your instructor check your work.
9. Mix the compound, using clean water, according to instructions on the kit.
10. Carefully pour the compound into the sealing fitting according to instructions on the kit.
11. Have your instructor check your work.
12. Clean up any excess compound or other residue.

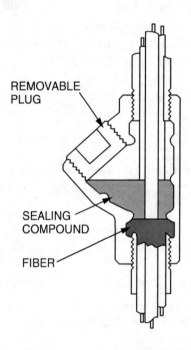

SUPPLEMENT

REMOVABLE
PLUG

SEALING
COMPOUND

FIBER

Figure 1 ■ Illustration of a Properly Filled Vertical Sealing Fitting

The NCCER makes every effort to keep these textbooks up-to-date and free of technical errors. We appreciate your help in this process. If you have an idea for improving this textbook, or if you find an error, a typographical mistake, or an inaccuracy in NCCER's Contren® textbooks, please write us, using this form or a photocopy. Be sure to include the exact module number, page number, a detailed description, and the correction, if applicable. Your input will be brought to the attention of the Technical Review Committee. Thank you for your assistance.

Instructors – If you found that additional materials were necessary in order to teach this module effectively, please let us know so that we may include them in the Equipment/Materials list in the Annotated Instructor's Guide.

Write: Product Development and Revision
National Center for Construction Education and Research
P.O. Box 141104, Gainesville, FL 32614-1104

Fax: 352-334-0932

E-mail: curriculum@nccer.org

Craft _____   Module Name _____

Copyright Date _____ Module Number _____ Page Number(s) _____

Description _____

_____

_____

_____

(Optional) Correction _____

_____

_____

(Optional) Your Name and Address _____

_____

_____